W9-AFH-993

SCIENCE

SCIENCE *and the* SOCIAL ORDER

by BERNARD BARBER

with a foreword by ROBERT K. MERTON

GP

GREENWOOD PRESS, PUBLISHERS
WESTPORT, CONNECTICUT

Library of Congress Cataloging in Publication Data

Barber, Bernard.
Science and the social order.

Reprint of the ed. published by Free Press, Glencoe, Ill.
Includes bibliographical references and index.
1. Science--Social aspects. I. Title.
[Q175.5.B36] 301.5 78-1569
ISBN 0-313-20356-3

Reprinted in 1978 by Greenwood Press, Inc.
51 Riverside Avenue, Westport, CT. 06880

Printed in the United States of America

10 9 8 7 6 5 4 3 2 1

FOR ELINOR

Contents

Acknowledgment

WITH great pleasure I report the debts I have acquired in the writing of this book. The two largest are to my friends and teachers, Talcott Parsons and Robert K. Merton. Only those who know the lectures and writings of Parsons and Merton will know how much I owe them for the substance as well as the sociological point of view of this book. For some fifteen years now, Merton's outstanding work in the sociology of science has been my model of what an enterprise in this field should be. I cannot count myself a "student" of George Sarton, but I have learned essential things from him about the history of science and about humanism during our long and pleasurable acquaintance. He has always been most generous in allowing me to use his personal library at Widener 189. Harry M. Johnson, by his minute criticism of the content and style of this book, has helped me very much. Several other friends have been kind enough to read part or all of the book in manuscript and to make suggestions for its improvement: Daniel Aaron, David Donald, Renee Fox, Frank H. Hankins, Alex Inkeles, Marion J. Levy, Jr., Charles H. Page, and Melvin Richter. Since I have not been able to use all their suggestions, these friends are not responsible for the shortcomings that remain. Finally, I have an immeasurably large debt to my wife, Elinor G. Barber, for her valuable criticism and support.

BERNARD BARBER

Dobbs Ferry, N. Y.
August, 1952

Foreword

ALTHOUGH the interaction between science and society has been a subject of occasional interest to scholars for more than a century, there has been little effort to provide a systematic organization of the facts and ideas which comprise that subject—the sociology of science. Numerous works, particularly in recent years, have variously dealt with one or another part of the subject—for example, the writings of Bernal, Crowther and Farrington, of Lilley, Pledge and Hogben. But these, with the important exception of Lilley's "Social Aspects of the History of Science," have not examined the linkage between science and social structure by means of a conceptual framework that has proved effective in other branches of sociology. It is the special distinction of this book that it puts into provisional order an accumulation of otherwise fragmentary and uncoordinated materials on the interplay of science and society.

When a book is clearly and closely organized, it becomes redundant to sketch out its design in a foreword. That is certainly the case with Mr. Barber's book. There is no need to enumerate its major themes here, for he does that himself, lucidly and succinctly. But there may be some value in attempting to place this book, and what it represents, in its social setting, to consider why it is that we have had to wait so long for a book which essays the task Mr. Barber has set himself: "to get a better understanding of science by applying to it the kind of sociological analysis that has proved fruitful when directed to many other kinds of social activities." How does it happen that the sociology of science is still a largely unfulfilled promise rather than a highly developed special field of knowledge, cultivated jointly by social, physical, and biological scientists? What are its present resources and prospects?

It is nothing new to observe that this field has long remained in a condition of remarkable neglect. In his recent diagnosis of "the present state of American sociology," for example, Edward Shils counts the study "of science and scientific institutions" among the major undeveloped areas of sociological inquiry. The evidence for such a judgment is varied but consistent. Consider for a moment the sphere of teaching: of the several thousand classes devoted to one or another branch of sociology in American colleges and universities, fewer than a handful are devoted to the sociology of science. Textbooks which, after an appreciable lag, ordinarily mirror the foci of attention in a discipline, also bear out this impression of neglect. Among current introductory textbooks in sociology, typically designed to acquaint students with the specialized spheres of interest in the field, all deal at length with the institutions of family, state, and economy, many with the institution of religion, but very few indeed with science as a major institution in modern society. Incidental discussions of the "important role" of science in society turn up in abundance, but there is little by way of a systematic analysis of that role.

Or consider the evidence in the sphere of research. It is true, of course, that comparatively little research is conducted in sociology altogether. In contrast to the tens of thousands of inquiries reported annually in physics, chemistry, and biology or, for that matter, of the thousands in history and English literature, there are only hundreds reported for the entire field of sociology. Among these, scores of studies deal with the sociology of marriage and the family, with population and with crime, and an appreciable number deal even with the sociology of religion, but the sociology of science has not yet enlisted sufficient interest to merit separate notice in the annual catalogues of sociological research.

Another telling sign of this studied neglect is found in the social organization of social research. Particularly, though not exclusively, in social science, institutes for specialized research are typically established in response to social, economic and political needs, as these are defined by influential groups in the society. Each "social problem" seems to generate its own complement of research centers. Thus, as public alarm is voiced at the alleged instability of the family and the rising rate of divorce, universities establish institutes specializing in research on the family; as the course of world affairs focusses attention on Russia, the Near East or the Far East, universities establish institutes devoted to social research on these regions. Yet among these scores of research centers in the social sciences, not one is devoted on any substantial scale to the sociology of science.

The inventory of neglect need not be continued. These varied evidences all reflect the central fact that the sociology of science claims the undivided attention of only a negligible number of specialists,—most of these in England and elsewhere in Europe. Among the several thousand American sociologists, not even a dozen report this as their field of primary interest. Indeed, the sociology of science has been nudged into being, not so much by sociologists as by occasional physical and biological scientists who occupy their leisure hours by working on this subject. That it is these scientists who have contributed most to what is presently known in the field can be seen from Mr. Barber's critically selected bibliography. Of the many books and articles on which he has drawn, roughly half were written by practicing natural scientists or by natural scientists who have turned to administration; more than a quarter, by historians and philosophers of science; and only the remaining fraction by sociologists. Granted that these numbers are rough approximations and that they may reflect the bent of the author who, for one reason or another, may have leaned toward writings by natural scientists. Yet other, more inclusive and less exacting, bibliographies of the sociology of science have much the same character: not many persons cultivate the field altogether and those who do are for the most part physical and biological, rather than social, scientists.

All this has left its mark on the nature of existing materials in the sociology of science. Since many of those interested in the relations between science and society are primarily engaged in other fields of inquiry, they have usually not been able to express this interest in the form of time-consuming disciplined research. They have, instead, written speculative books and articles, making use of the historical evidence lying close to hand. In these works, therefore, the historical anecdote often stands in place of systematic data and the opinion, in place of documented inference. Generalizations spring easily from a few selected particulars. Thus, Newton is noted to have been a bachelor and so, celibacy manifestly contributes to a wholehearted devotion to science; in his invasion of Egypt, Napoleon was accompanied by nearly two hundred astronomers, archaeologists, chemists, geometers and mineralogists, from which it appears that war generally facilitates scientific advance. These works also draw, again and again, upon the same small number of empirical studies, although these are insufficient to bear the weight of the numerous conclusions which are precariously based upon them.

The circumstance that much of the material in this field has been fashioned by physical and biological scientists for whom this is an avocation rather than a major concern has left another mark, which Mr. Barber has tried to erase. Unlike the pattern in solidly established disciplines, in the sociology of science facts are typically divorced from systematic theory. Empirical observation and hypothesis do not provide mutual assistance. Not having that direct bearing on a body of theory which makes for cumulative knowledge, the empirical studies that have been made, from time to time, by natural scientists have resulted in a thin scattering of unconnected findings rather than a chain of closely linked findings.

As a consequence of all this, the sociology of science has long been in a disordered condition: on the one hand, it is unduly speculative, with few established facts altogether and, on the other, it suffers from an excess of empiricism, since these facts are ordinarily not cast in the mould of theory. Largely absent in this field are the productive patterns of inquiry in which, as has been said, men pursue facts until they uncover ideas or pursue ideas until they uncover facts.

It is by no means inherent in the subject-matter that the years and decades should have slipped by with relatively few accretions to our knowledge of the interconnections of science and society. Rather, this is the natural outcome of continued neglect. Since the social scientists having the needed grounding in theory have not ordinarily taken up empirical studies in the sociology of science and since the physical and biological scientists who have conducted such studies ordinarily lack the needed theory, it is no wonder that the growth of the field has been stunted. Not that mere numbers of students devoted to a special branch of knowledge ensure its rapid growth—some problems remain refractory to quick solution—but the converse is a truism: a field of knowledge does not prosper if it is neglected.

The slow, uncertain and sporadic development of the sociology of science has meant that its leading ideas have grown worn with repetition. As one example among many of this, consider the history of the inferences that have been drawn from the multiple and independent appearance of the same scientific discovery, or of the same invention. It may not be too much to say that the implications about the cultural context of innovation which have been drawn from this strategic fact are among the more significant conceptions found in the sociology of science. These conceptions are, properly enough, associated with the sociologists, Willim F. Ogburn and Dorothy S. Thomas, who drew up a list of almost 150 independently duplicated scientific discoveries and technological inventions, pointing out that these innovations became virtually inevitable as certain types of knowledge accumulated in the cultural heritage and as social needs directed attention to particular problems.

In two respects, the history of this idea illustrates the slow pace of development in the sociology of science: first, this idea has been little elaborated or extended since it was emphasized by Ogburn and Thomas a generation ago and second, essentially the same idea regarding the sociological significance of multiple independent discoveries had been repeatedly formulated, particularly throughout the century before. As early as 1828, Macaulay, in his essay on Dryden, had noted that the independent invention of the calculus by Newton and Leibniz belonged to a larger class of instances in which the same discoveries and inventions had been made by scientists working apart from one another. This coincidence he attributed to an accumulated common stock of knowledge and to a common focus of attention. As he put it, in terms now grown old with repetition, "Mathematical science, indeed, had then reached such a point, that if neither [Leibniz nor Newton] had ever existed, the principle must inevitably have occurred to some person within a few years." Nor was Macaulay alone in this idea, which turned up in the most diverse quarters of English society. In spite of Carlyle's doctrine of culture heroes, this unheroic idea was regarded as a useful commonplace by the Victorian manufacturers who testified before Royal Commissions that, after all, inventions only constituted small and inevitable increments in existing technology as could be seen from repeated instances of the virtually simultaneous and independent appearance of the same invention. Not much later, a prominent writer who detested his own position as a Manchester manufacturer expressed the same thesis when he wrote of his partner in ideas that "while Marx discovered the materialist conception of history, Thierry, Mignet, Guizot and all the English historians up to 1850 are the proof that it was being striven for, and the discovery of the same conception by Morgan proves that the time was ripe for it and that indeed it *had* to be discovered." Meanwhile, the same conception, based on the same kind of evidence, was being promulgated in the United States. In 1885, William H. Babcock and P. B. Pierce were informing their colleagues in the Anthropological Society of Washington that the "synchronism of inventions" testifies that "the progress of a certain art has reached a point where a given step becomes inevitable" and that "this shows that the individual man is of less importance in invention than his environment." Shortly thereafter in France, Gabriel Tarde, in 1902, and Abel Rey, in 1922, were arriving at the same conclusion, observing that the simultaneity of discoveries and inventions was sufficient evidence of the crucial part played by cultural accumulation.

The point of this is not, of course, that Macaulay or the Victorian manufacturers, that Engels or the American anthropologists said it first. Nor that this multiple and, in some instances, independent rediscovery of the same idea may be regarded as an hypothesis which is confirmed by its own history. Nor again, is it to detract from the genuine contribution of Ogburn and Thomas who did so much to establish this hypothesis in sociological thinking. The point is, rather, that periodic rediscovery of the same hypothesis characteristically results from the neglect, by sociologists, of the sociology of science and technology so that this special field, in contrast to other specialized branches of sociology, has developed little that is new through the years. Very few, for example, have followed up the implications of the hypothesis to determine, through actual empirical study, the extent to which, as the hypothesis supposes, the same constituents were indeed comparably developed in the different cultures where the same discovery or invention appeared. Consequently, this hypothesis, like other hypotheses in the sociology of science, has remained substantially unextended for more than a century.

It is no easy matter to say why it is that the sociology of science has for so long remained in a state of comparative lethargy. This condition is particularly anomalous, since it seems widely agreed that science constitutes one of the major dynamic forces in modern society. Possibly there are social and institutional circumstances which, largely unnoticed, combine to divert the attention of scholars and scientists from a subject which one would expect to be of central interest in a world where science looms large.

The relative neglect of this subject by physical and biological scientists perhaps requires little explanation. After all, specialization in science calls for devoted concentration of effort, and the sociology of science is not their *métier*. Hard at work on research in their own science, they are scarcely in a position to take up yet another life as sociologists. Furthermore, current practices and assumptions in the world of natural science may militate against their developing even a casual interest in the linkages of science and social structure. For example, there may prevail, among these scientists, the assumption that the history of science is comprised by a succession of great minds—an assumption with an easy plausibility since turning-points in the history of science are indeed commonly associated with great scientists. Standing on such an assumption, scientists may readily lose sight of the less visible social processes which play their indispensable part. In paying its homage to these great minds, society may inadvertently reinforce that assumption. Eponymy, the practice of affixing a scientist's name to his discovery, as with Boyle's law or Planck's constant; Nobel prizes and lesser testimonials; nationalistic claims to scientific preeminence which lead to a focus on the contributions of one's own nationals; the virtual anonymity of the lesser breed of scientists whose work is indispensable for the accumulation of scientific knowledge—these and similar circumstances may all reinforce an emphasis on the great men of science and a neglect of the social and cultural contexts which have significantly aided or hindered their achievements.

Physical and biological scientists may be reluctant to consider the bearing of social environment upon science for quite another set of reasons. They may be apprehensive that the dignity or integrity of their work might be damaged were they to recognize the implications of the fact that, as Mr. Barber points out, science is an organized social activity, that it presupposes support by society, that the measure of this support and the types of scientific work for which it is given differ in different social structures, and that the directions of scientific advance may be appreciably affected by all this. Or perhaps their reluctance comes from the widespread and mistaken belief that to trace such connections between science and society is to impugn the motives of scientists. But, as Mr. Barber and others have shown, this belief involves a confusion between the motives of scientists and the social environment which affects the course of science. It assumes also that scientists are consistently aware of the social influences which affect their behavior, and this is by no means a self-evident truth. To consider how and how far various social structures canalize the directions of scientific research is not to arraign scientists for their motives. Nor, as Mr. Barber reminds us by emphasizing the relative autonomy of science, is it to make the institution of science the mere appendage of political, economic, and other social institutions.

Whether these are, or are not, the reasons why physical and biological scientists neglect the sociology of science, they can scarcely be the reasons why sociologists have given this field such little attention. The mythopoeic view of history has had little standing among them for generations—if anything, they are more likely to underestimate the distinctive role of great men in social change. Nor do sociologists generally assume that to study the social patterning of human behavior is to condemn the motives of those acting out those patterns—they are more likely to adopt the relativistic opinion that to understand is to excuse, that the conception of individual responsibility is alien to social determinism. It would seem, therefore, that the absence of concerted research interest in this field among sociologists must have another explanation.

Although there is little evidence on which to base an explanation, the fact itself is at once so conspicuous, and strange, that it invites conjecture. It may be that the connections between science and society constitute a subject-matter which has become tarnished for academic sociologists who know that it is close to the heart of Marxist sociology. Such an attitude need not stem from a fear of guilt by association with politically condemned ideas, though this, too, may play a part. Like attitudes toward most revolutionary ideas, attitudes toward Marxism have long been polarized: they have typically called for total acceptance or for total rejection. Sociologists who have come to reject the Marxist conceptions out of hand have not uncommonly rejected also the subject-matters to which they pertained: American sociologists do not much study the conflicts between social classes just as they do not much study the relations between science and society. At the other pole, those who regard themselves as disciples of Marxist theory seem to act as disciples merely, content to reiterate what the masters have said or to illustrate old conclusions with newly selected examples, rather than to consider these conclusions as hypotheses which they are to test, extend, or otherwise modify through actual empirical inquiry. At both polar extremes, the sociology of science suffers, either by inattention or by preconception.

In part, also, the field is the victim of existing programs of higher education. Physical and biological scientists have typically had their rigorous training confined to the specialized skills and knowledge of their field, and few have had more than a slight acquaintance with social science. Social scientists, similarly, have typically had little training in one or another branch of the more exact sciences or even in the history of science, and consequently feel reluctant to take up a specialization for which they see themselves as unprepared. In the meantime, the sociology of science falls unnoticed between these two academic stools.

Yet to emphasize the relative neglect of this field is not to say that it is wholly barren or condemned to slow growth. Mr. Barber's book would belie any such rash claim. Actually, there are many signs that this condition of neglect is drawing to an end and that the prospects for growth are greater than ever before.

Various social tendencies, not entirely new but now more conspicuous and compelling, are forcing attention to the relations between science and its environing social structure. The politicalizing of science in Nazi Germany and in Soviet Russia, for instance, has aroused the interest of many in identifying the particular kinds of social contexts in which science thrives, a problem central to the sociology of science and one which Mr. Barber treats more systematically than has been done before. In liberal societies as well, recent changes have subjected scientists to abrupt conflicts between their several social roles and between their deepseated values. Early in their apprenticeship, scientists have commonly acquired certain values which, as a result of changed social conditions, they are asked to unlearn and abandon at a later point in their career. The value, for example, which calls for making new-won knowledge part of the commonwealth of science now clashes with the demands made upon them, in their role as citizens, to keep some of this knowledge secret. Men previously unaware of the social contexts of their attitudes and values are apt to become acutely aware of them when they are frustrated in their aims by strains and stresses which are manifestly social in origin. Even the most artless and singleminded of scientists, living out their work-lives within the confines of the laboratory, must now know, to adapt a remark by Butterfield, that they are "not autonomous god-like creatures acting in a world of unconditioned freedom."

More particularly, these historical developments have given rise, among scientists themselves, to polemics and controversies about the "social control of science,"—a hot conflict which Mr. Barber judiciously analyzes in Chapter X of his book. However unproductive they may be in settling the points at issue, these warring opinions have had the collateral result of exciting and maintaining interest in the social relations of science at a higher pitch than ever before.

Not only scientists, but a wider public, have had their attention drawn to the social implications of science by recent events. The explosion over Hiroshima, and other experimental atomic explosions, have had the incidental consequence of awakening a dormant public concern with science. Many people who had simply taken Science for granted, except when they occasionally marvelled at the Wonders of Science, have become alarmed and dismayed by these demonstrations of human destructiveness. Science has become a "social problem," like war, or the perennial decline of the family, or the periodic event of economic depressions.

Now, as we have noted, when something is widely defined as a social problem in modern Western society, it becomes a proper object for study. Particularly in American sociology, new special branches have developed in response to new sets of problems. A few generations ago, the great influx of immigrants evoked deep sociological interest in the processes of assimilation and acculturation, just as changes in the status of Negroes in American society have intensified the specialized study of race relations. So, too, the more conspicuous problems of city life so nearly usurped the attention of sociologists that, for a time, their research-sites were typically in urban slums, the better to observe the behavior of juvenile delinquents, adult criminals and other presumed aberrants. With the great diffusion of the motion picture and the appearance of radio, there began a new phase of concerted research on mass communications and public opinion, thus reviving another sociological specialty on a scale previously unknown. In more recent years, the effective organization of trade unions in this country and the attendant organization of conflict between workers and employers belatedly brought in their wake the special field of industrial sociology.

There are indications that the sociology of science, as a distinct field of specialized research, is now in much the same situation as was industrial sociology a scant twenty years ago. Previously amorphous and sporadic interest in the subject is becoming crystallized and continued. There is, however, a basic difference in the social contexts of the two fields which may make for a different outcome: industrial sociology concerned itself largely with problems involving the economic interests of industry—with problems of worker morale, with the connections between informal group structure and productivity, with relations between management and labor. As with technological research, when it promised rich yields, so with sociological research: industry was prepared to support these studies because it was good business to do so. Profit-making organizations are constrained to make their decisions in terms of expectable profit, and, in this narrowly economic sense, sociological studies of science and the scientist hold little promise. It is from the agencies not organized for economic gain that support must come.

Out of the complex of recent historical developments—among them, the attempts to subordinate science to political control, the deepening conflicts between men's roles as scientist and as citizen, the events leading science to be widely regarded as a source of social problems—there has come a renascence of interest in the sociology of science. Thus, in cooperation with the American Academy of Arts and Sciences, Phillip Frank has lately gathered together a group of scholars to carry forward empirical and theoretical studies in this field. Another group has been formed, under the auspices of the American Council of Learned Societies, to study the humanistic, including the social, aspects of science. *L'Union International d'Histoire des Sciences* has broadened its scope to include a Commission for the history of the social relations of science and its important first report, prepared by S. Lilley, gives ample proof of its sociological orientation. Some of the relatively few university departments of the history of science have also begun to attend to sociological considerations, and this may be expected to develop all the more rapidly as suitable research materials accumulate.

Another kind of academic development promises, in due course, to provide some of these research materials. For more than a decade, sociologists have exhibited increasing interest in the structure, role, and functions of the professions in society—medicine and the law, the ministry and engineering, among others, are being studied in their social implications. This may well carry over to comparable studies of science and scientists. Should this happen, it would have the further advantage of making for a synthesis of historical materials and of materials based on first-hand field work. Until now, the great bulk of studies in the sociology of science have been based almost entirely on historical data—the documents left behind by scientists, autobiographies, diaries, and reports of scientific societies. This is of course indispensable material, but it is not sufficient. Scientists, like others, are apt to be so deeply immersed in their own work that they cannot take cognizance of the multitude of social actions and interactions which presumably take place in the laboratory, as in the factory, and which are, in significant degree, below the threshold of awareness of those involved in them. There already exists, of course, a vast literature on "scientific method" and, by inference, on the "attitudes" and "values" of scientists. But this literature is concerned with what the social scientists would call ideal patterns, that is, with ways in which scientists *ought* to think, feel, and act. It does not necessarily describe, in needed detail, the ways in which scientists actually do think, feel, and act. Of these actual patterns, there has been little systematic study—the psychological examination of biologists and physicists by Roe representing a rare exception. It is at least possible that if social scientists were to begin observations in the laboratories and field stations of physical and biological scientists, more might be learned, in a comparatively few years, about the psychology and sociology of science than in all the years that have gone before.

From all this, it seems that this book could scarcely have appeared at a more fitting time. For at a time of renewed interest in a field, even a single book, which provides a tentative systematic overview of that field, can have a disproportionately great effect. It is not unlikely that Mr. Barber's book, together with others that will probably follow in its path, will do much to encourage the establishment of university courses introducing students to the sociology of science. It may well be that some of the students electing these courses, perhaps because the recent course of history has aroused their curiosity about the social environment of science, will develop an abiding interest in the subject. These would then be the new and substantially the first generation of recruits, trained both in social science and in one or another of the physical and biological sciences, who, as they mature into independent scholars, could establish the sociology of science as a specialized field of disciplined knowledge. Mr. Barber's book takes a long step in that direction.

ROBERT K. MERTON

SCIENCE AND THE SOCIAL ORDER

INTRODUCTION

What Is Science?

BECAUSE we can ask this simple question, many of us expect a simple answer for it. But in fact we need a complex answer to tell us the many different things that science has been and is, or, to put it another way, the several aspects that science has. There is, of course, a certain unity and integration in science, a unity which is something less than complete, to be sure, but which is nevertheless an important condition of its existence. We shall have much to say of this unity later on. But science also has many separate aspects. We shall find that a satisfactory understanding of science requires the isolation and study of these several aspects of science as much as it does the careful study of the unity itself.

We have only to examine the multiplicity of public and private images there are of science to see how many different facets it presents. Science is a man in a white coat, most often, probably bending over test tubes in a laboratory. Or science is Einstein's theory of relativity, known by a formularistic tag, $E=mc^2$. A complicated machine, perhaps one of the new electronic mathematical computing machines, described by some writers as "mechanical brains," is still another common symbol of science. In the Great Depression of the 1930's, science meant technological unemployment to many, a Frankenstein's monster turned on its own creator, Society. More often, despite the atomic bomb, science still means the fulfillment of hope and the persistence of hope—it discovers insulin, penicillin, perhaps even a remedy for that trivial scourge, the common cold; it is always enlarging our material welfare; and it never stops seeking a cure for cancer, for poliomyelitis, for psychosis, and for myriad other ailments of mankind.

Science shows all these aspects and a great many more. Each of us can multiply his own conceptions of science. But we need a systematic understanding of science, a way of relating this diversity of its nature to its underlying integrated unity. Science is no randomly collected assortment of elements and activities, but a coherent structure in which the parts have functionally interdependent relations. In short, we need a more scientific understanding of science itself.

One way, an obvious but somewhat neglected way, of gaining this systematic comprehension of science is to consider it first and fundamentally as a social activity, as a set of behaviors taking place in human society. In this perspective, science is more than disembodied items of guaranteed knowledge and more than a set of logical procedures for achieving such knowledge. In this perspective, science is, first of all, a special kind of thought and behavior which is realized in different ways and degrees in different historical societies. We often take our own society and our own science for granted, as if they were universal in just their present form. We do not see that other societies have dealt quite differently with rational thought and activity, which are the essence of science; we do not see that our own great approval of science is historically unique.

By taking science systematically as a social activity, we can perhaps see the determinate connections that it has with the different parts of a society, for example, with political authority, with the occupational system, with the structure of class stratification, and with cultural ideals and values. And because these political and occupational and class and cultural systems vary among societies, we can see that some societies—notably our own—are much more compatible with science than are others, and also we can seek out the social sources of this variable compatibility. We can, for example, see how certain great and interwoven changes in a society, such as those that occur in revolution or war or economic depression, affect both the rate and the direction of growth in science.

This social approach permits us to make yet further progress in deepening our understanding of scientific activity. Science is carried on, we shall see, in different kinds of social organizations in different societies. In our society, it is almost wholly carried on in universities and colleges, in industry and business, and in Government groups. Science had a different social locus in Greek society.

Each of the different types of social organization performs different functions for science and presents characteristic problems. Analysis of these several social contexts of scientific activity should provide us with a better understanding of the nature of science.

Still further. When we take this view of science as a social activity, we can see how its products—its inventions and discoveries—are the products of a process that has essential social characteristics. Here are some of the questions that need to be asked and answered in the light shed by seeing science as an irreducibly social activity: —Is necessity the mother of invention, as is so often asserted? How does society define "necessity?" Do inventions occur by chance? in clusters? What is the relation between the individual and society in the process of discovery?

And finally, seeing science as a social activity can direct our attention more fruitfully to some of the "social problems" of science, to the problems of the social control of science. Not only is science in part dependent upon its supporting society, but also it acts in part independently upon that same society. This is only to say what everyone now knows and cannot blink, that science has social consequences. To mention only a few of these, science changes the structure of the economy, it seems to challenge established religion, and it shakes up the relations among the members of families and communities. We develop mixed feelings as a result of these and other social consequences; because we think some of them good, others bad, our attitudes toward science become confused or ambivalent. Then scientists are challenged to justify their ideals and their activities. Indeed, scientists and laymen alike discuss with considered seriousness "the social responsibilities of science." Some who discuss this problem propose that science be "planned," some scientists and some laymen; whereas others in both camps have very strong objections to "planning." We need to see in what sense science as a social activity can be "planned," in what degree, not.

This then is the approach we shall take to the study of science, these are some of the questions we intend to put and to answer at least tentatively. Our task is to get a better understanding of science by applying to it the kind of sociological analysis that has proved fruitful when directed to many other kinds of social activities. This is not, of course, a task in which we have to start from scratch.

Many scientists and many students of the history of science have noted, with greater or less clarity, that science is a social activity. The results of their observation and analysis are freely available and have been used throughout this study. What we wish to build upon these foundations is a fully explicit and systematic sociological analysis of science, so far as that is now possible with current general social science and current factual knowledge about science. The structure of understanding here erected will serve its purpose if it can be used as a foundation for further progress in this area of sociological study and as an instrument in dealing with the practical social problems of science-in-society.

I

The Nature of Science: The Place of Rationality in Human Society

MAN HAS always dreamed of, but never actually lived in a Garden of Eden. It is of the essence of the human condition that man lives not in a compliant but in a resistant environment, an environment which he must constantly make an effort to control, if he cannot wholly master it. Man's physical and social situations are ever setting tasks for him in which he must somehow efficiently adapt means to ends. For if it is inherent in man's situation to have to expend "effort" to cope with the environment, it is also in his nature to have a limited amount of energy for this general effort. Man everywhere and at all times, therefore, has had to make at least some of this effort efficiently and economically.

In his need to economize energy, in his need to adapt means to ends efficiently, man has always had the indispensable aid of his power of rationality and of some knowledge about his environment. In our next chapter we shall give historical evidence of this fact from a variety of societies widely scattered in space and in time. Here it is enough to recognize the universality of human rationality, to examine its characteristics somewhat more closely, and to show its connection with science. For this is the essential point from which our whole investigation starts: that the germ of science in human society lies in man's aboriginal and unceasing attempt to understand and control the world in which he lives by the use of rational thought and activity. I take it that Professor Percy Bridgman, the Nobel Prize physicist, was making much the same point when he said, "I like to say there is no scientific method as such, but rather only the free and utmost use of intelligence."[1] We shall see, of course, how rationality and intelligence must be disciplined before they become the highly developed science we are familiar with, but it is essential to understand first this prime human source of science.

Let us look a little more closely, then, at what we mean by human rationality. In its most general sense, by "rational thought" we mean simply any thought which is in accord with the canons of Aristotelian logic, or, for certain cases, with modern, non-Aristotelian logic.[2] We mean, for example, that rational thought keeps non-identical things separate (A cannot be both A and non-A) and that it follows the processes of syllogistic reasoning about the connections between things. Thought may be rational in this fashion whether men who use it are explicitly aware of these logical canons or not. Thus there was rational human thought, in which logical rules of reasoning were implicit and effective, long before Aristotle made his brilliant formal statement of the rules. And also, a great deal of rational thought since Aristotle and even today proceeds without any self-conscious use of formalized logical rules. Whether they use logic explicitly or only implicitly, all men are in some measure potentially capable of rational thought and activity and use them in their everyday lives.

It is well to be clear that our definition excludes certain kinds of thought which are sometimes also labelled "rational." We include only those which conform to the rules of logic and not those which are in accord with other kinds of norms and canons of relationship. Thus we exclude thought which may be called "rational" because it follows certain moral norms or certain norms of esthetics. These canons of beauty, of taste, and of ethics may indeed be "rational," in terms of their consistent relations one with another, but this is another sense than the one intended here. We mention these other canons of relationship because, like the canons of logic, they occur not only in our own society but in all other societies, though of course they vary in substance. In all societies these several types of relationship must be kept separate, and it is perhaps significant of the power of logical canons and of scientific rationality in our own society that we are constantly trying to extend both of them into areas where other norms of relationship are relevant. Perhaps it is because we are so much impressed with science that we sometimes speak of a "rational art" and a "scientific ethics." In any case, here and now we are interested only in that rationality defined by logical norms, for it is from these that science springs.

We must next recognize that logical rationality does not have any one-to-one relationship with science and that it is therefore the source of much else besides science. This is to say that rational thought as we have defined it may be applied to different kinds of goals which occur in society and that science exists only when rational thought is applied to one of these kinds of human ends. It is not science when men talk about the existence of God or the nature of Evil, although their discussion of these subjects may be carried almost up to its final point in full rational accord with the rules of logic. Science exists only when rational thought is applied to what we may call "empirical" ends, that is, ends which are available to our several senses or to the refined developments of those several senses in the form of scientific instruments. It is at once apparent that there is a vast range of such empirical ends in society. A very large number of such ends are summed up, in every society, under the general empirical goal we refer to as "the control of nature." It is an empirical end of every society to achieve a sufficient control of nature to make agriculture and industry reliable undertakings. And therefore science is applicable to industry and agriculture insofar as men expend their energies in these activities in accord with the rules of logic. In sum, science must be both rational and empirical.

It is beyond our present purposes to define the other kind of ends, the non-empirical ends, other than residually, as ends about which we may and do use rational thought but which are not available to our sensory equipment and its instrumental extensions. We need simply note that such ends always do occur in society; in short, that there is always social thought in connection with such matters as salvation, good and evil, justice, and the like. Indeed, as anyone can see who reads the great religious thinkers of any society, the application of logical thought in these matters has reached a very high level of development. The philosophy and theology of the scholastic thought of the Mediaeval period in Western society is only one notable example of such a development of non-empirical rationality.

In all societies, then, rational thought is applied to both kinds of end, the empirical and the non-empirical. But it is a very important fact that the degree of interest in these two different kinds of end varies widely among different societies. We shall look into this variation in our next chapter. We shall see, for example, that our own modern Western society has a uniquely large, though far from exclusive, concern for empirical as against non-empirical ends. Hindu society has not shared this relatively greater emphasis on the empirical; it has had a relatively greater interest in the non-empirical than does modern Western society.

Rational thought about either one of these two kinds of end is not, of course, without influence on rational thought about the other kind. Thus, although it is from the application of rational thought to empirical ends that science derives directly, the development of rational thought about non-empirical ends has had an indirect connection with the evolution of science. Religious rationalism, for example, provides skills in rational thinking which have often been diverted later to empirical ends. On this view science is the indirect heir of the great achievements of skill in the rational treatment of non-empirical subjects made by Greek and Mediaeval Western societies.

It may seem that we have now drawn a universal, fixed, and sharp line between empirical ends and non-empirical ends. Such a line does not exist. The ends which are defined as empirical vary somewhat from society to society. For instance, many human ends which the science of biology and its medical applications in our society treat as empirical are in other societies not so treated. That is to say, we consider disturbances of health as suitable objects for the application of scientific techniques of diagnosis and treatment. In many other societies, a great many diseases which we consider sickness have been presumed to be rather the effect of offenses against some supernatural, non-empirical powers and therefore not at all a proper object for empirical techniques of curing. The scope for science varies with the realm which is conceived as empirical. Moreover, not only does the area of what is considered empirical vary within important limits among different societies, but also its size changes historically within any given society. Even in our own society, the problem of physical and mental health has only slowly come more and more into the realm of the empirical. Social problems have been among the last to gain empirical status, and only recently have we come to treat alcoholism, to take only one such example, as partly a matter of science rather than entirely a matter of sin.

To say that the realm of the empirical varies somewhat among societies and somewhat in time is not, however, to assert that *all* non-empirical ends in society are potentially reducible to the empirical. Such a reducibility has indeed been held to be possible by some extreme versions of empiricist philosophies. The course of recent history has seemed to some of us to be bearing out these philosophies, to be reducing constantly through science the area of the non-empirical. As science has advanced steadily into certain areas that were formerly considered non-empirical, some of us have assumed that eventually only science would remain and no non-empirical problems. But this view has come to seem less tenable even to many who formerly held it, and, indeed, there has been an important development in recent social science which asserts quite the contrary position.[3] The newer assumption is that the non-empirical entities referred to by social values, religious ideas, and social ideologies quite positively do have a necessary and independent status. They are, of course, affected by scientific ideas, just as they in turn affect scientific ideas, as we shall see, for example, when we discuss the influence of religious ideas on the rise of modern science. But they have a margin of autonomy and are not wholly reducible now or ultimately to proper empirical science. Since this is so, science alone can never provide for man a complete adjustment to the natural and social world. We shall have many occasions to see the dependence of science upon certain ultimate, non-empirical social values and world-views. Yet, there is a certain impiricist philosophy which still believes that science *is* all-sufficient for human adjustment, and we shall have to consider this fallacy again in our discussion of the social consequences of science.

We have up to this point in our discussion been concerned to locate the origin of science in empirical rationality. But there are other problems than those of origin; there are problems of relative development. Not all valid rational thought about empirical ends is of the same degree of development; all such thought is not science as we know it in the modern Western world. While every known society necessarily has considerable rational knowledge of its empirical environment, based in what we may call "common sense" and empirical lore, not all societies have the highly developed scientific theories which exist in our society. The forms of empirical rationality in society are many, that is, and have evolved over the course of history. Relatively undeveloped forms, what we may consider relatively undeveloped science, are limited to particular empirical situations and to fairly specific empirical ends. Such science is "not detached from the craft;" it is rule-of-thumb rationality; it is not highly general or systematic. It is, for example, a lore of curing, like that which exists in many non-literate societies, and not a highly generalized science of chemistry and biology, such as forms the basis of modern medical therapy. Relatively highly developed forms of empirical rationality, by contrast, like those which make up the essence of modern science, are extremely general and systematic sets of ideas. Such science abstracts from a near infinity of particular situations. It is, for example, a Newtonian or Einsteinian theory of the whole universe, expressed in a few general ideas, not a craft of weather prediction or a lore of astrology. Nevertheless the less developed and the more developed forms of science have a common origin and the latter have evolved out of the former. In our next chapter we shall trace the historical course of this development of science in some detail, trying also to show how different social factors have influenced it directly and indirectly. But before we can take up that task, we need to consider more closely than we yet have the nature of highly developed science. Here we have a set of problems the understanding of which is an essential preliminary to our analysis of the social aspects of science.

President Conant of Harvard, who is himself a chemist, has recently described the essential functions for all science of those highly generalized and systematic sets of ideas which we have just said are the heart of highly developed modern science.[4] He calls these ideas "conceptual schemes." It is with the nature of conceptual schemes and their relations with such matters as experimentation, mathematics, and "common sense" that we now wish to deal.

In a formal definition, conceptual schemes might be said to be more and less general systems of abstract propositions of empirical reference which state the determinate conditions under which empirical phenomena are related among themselves. By "related among themselves" we mean *both* remain constant *and* change. Science has learned that only if it can know the conditions under which things change can it satisfactorily explain also why they don't change. Without adequate conceptual schemes, scientific research is either blind or fruitless. President Conant has demonstrated this essential fact about science with examples from the history of science, especially of the seventeenth and eighteenth centuries. For example, what Aristotle and Galileo following him could not do with the theory of the vacuum and with the air pump because of their inadequate view that "nature abhors a vacuum," Torricelli and Robert Boyle could do when they devised a more adequate conceptual scheme for the same phenomena based on the notions that air has weight and is an elastic medium. Or similarly, Lavoisier in the eighteenth century laid one of the important foundations for modern chemistry when he abandoned the ancient phlogiston theory to explain the process of burning and replaced it with a more adequate conceptual scheme about oxidation and combustion. Further examples could be multiplied from the history of science. Indeed, the history of science, and especially of modern science because of its rapid rate of progress, could be written in terms of the successively greater development of conceptual schemes and of the correspondingly greater reduction in the degree of empiricism in science. Good conceptual schemes, says President Conant in sum, are the essential *cumulative* component of all science.

For reasons not only of adequacy but also of parsimony and elegance of thought, the ideal conceptual scheme at any given time is that which has the greatest generality, that is, the one in which the number of conceptual categories or variables in terms of which abstract general propositions are stated is very small. The greatest ideal of this kind yet achieved in science exists in the physical sciences, for example, in the Newtonian conceptual scheme, which is constructed in terms of such extremely few and general variables as mass, force, motion. In the same sciences, the Einsteinian and quantum mechanics re-formulations of the Newtonian theories seem to be even more general conceptual schemes. Unfortunately not all scientific conceptual schemes are anywhere near so general and systematic as those of the physical sciences, although what these sciences have accomplished is the ideal toward which all others strive. At least one other of the natural sciences, namely, biology, has not yet achieved a conceptual scheme of very high generality like that of the physical sciences. Therefore it is less adequate as a science. As for the social sciences, they tend to be still in a quite empiricist condition, with few if any general conceptual schemes that are widely accepted among professional workers in these fields. But of this we shall say more when we come to discuss the nature and prospects of the social sciences.

An understanding of the essential functions of conceptual schemes in science explains a certain paradox that has struck the attention of many students of the history of science. This is the paradox that mediocre minds and even the untrained minds of students in school often find it easy to understand things which baffled the minds of some of the greatest scientific intellects for centuries.[5] Things which elementary physics teachers find it easy to communicate to high school students, things which seem to everyone to be the obvious and natural way of regarding the universe, the obvious way of considering the behavior of falling bodies, for example, these things perplexed such great intellects as Leonardo da Vinci and even Galileo, when, as Butterfield puts it, "their minds were wrestling on the very frontiers of human thought with these very problems."[6] The point is that conceptual schemes always appear deceptively simple *after* they have been made and accepted. This is, of course, also an essential virtue, for otherwise they would not provide continual foundations upon which their successors could be constructed.

Despite the common and essential functions which they all perform, conceptual schemes may also vary in type, as well as in generality and systematization. Perhaps "type" implies a greater difference than actually exists, but certainly some differences must be noted. Probably the essential difference consists in the degree of precision or determinateness with which statements of relationship can be made. This difference occurs, for example, between the conceptual schemes of the physical sciences and those of the biological sciences. In the physical sciences, a much higher degree of precision and determinateness is possible because the empirical data to which their general abstract variables like mass and force refer can all be ordered to these variables in precise *metrical* terms. These data constitute genuine mathematical series, conforming to technical logical criteria like transitivity, etc.[7] As a result, for any given concrete system to which the conceptual schemes of the physical sciences apply, a precise system of differential equations can be formulated. These equations both describe the present state of the system and make it possible to derive determinate statements about changes which any part of the system will undergo as a result of even minute changes in any other part of the system. To take what is perhaps a stereotyped but clear example, Boyle's Law of Gases is a simple version of such a precise statement of variation in a concrete system to which the concepts of pressure, volume, and temperature are relevant. A similar but much less familiar example to the novice in science would be something like the Second Law of Thermodynamics, which states: When free interchange of radiant energy or heat conduction takes place between two bodies at different temperatures, it is always the hotter of the two that loses energy and the colder that gains energy.

Although many of the laymen among us may think the opposite, nothing like this degree of generality, precision, and determinateness is yet possible in the biological sciences. Biological analysis proceeds not in terms of differential equations, but rather mostly in terms of structural-functional analysis. That is, biology has still to content itself with describing first the *structural* components of its concrete systems and then with describing the *functions* of the processes of these systems. It does this by showing the contribution which these processes make to the maintenance of the stability and constancy of the structure of the system. For example, in order that a human organism may maintain what the physiologist W. B. Cannon has called "body homeostasis," or constancy of the structure of the organism, the supply of oxygen to the cells must be maintained. This is to say that the function of the respiratory and circulatory processes in the human body is, among other functions, to maintain this vital oxygen supply. However the processes of this function cannot be described in any very precise metrical form, as one may read in Cannon's fascinating book, *The Wisdom of the Body.*[8] And the same is true also, he shows, for a great many other functional needs of the body system. To take a few other examples, this is so too for the need for constancy within certain imprecise limits of the salt content of the blood, of blood sugar levels, and of body temperature. In consequence, as Ellice McDonald, Director of the Biochemical Research Foundation says, "Biological research in general is upon the experiment and error basis."[9] We may take it that this is an only slightly exaggerated way of saying that, relative to the physical sciences, biology is still in an empiricist condition.

Yet for all its relative lack of metrical precision and determinateness—and of course it is important to see that this is a relative lack—biology is a respected science which has a great range of useful application in medical and other technologies. It is important for the understanding, as well as for the future advance of, certain branches of science to see that rational knowledge may be fairly highly developed without being identical in form to the physical sciences. Holding the opposite view is an error which is not foreign even to some scientists.[10]

Our preceding discussion of "types" of conceptual schemes should not be taken to mean, of course, that the use of mathematics, in any fashion, is the essential difference between the physical sciences and the biological and other sciences. The other sciences also try to order their data in metrical series whenever they can, but this is much less often possible. Indeed, even in biological research, it is sometimes asserted, there is a fruitless straining after the use of mathematics. The pressure toward quantification comes from the higher prestige which the mathematical forms of the physical sciences have for many scientists. But Professor Cannon has asserted that "such intellectual snobbishness" is not justifiable so long as there are many important fields of investigation to which mathematics is not applicable. "The biologist," he says, "should not be looked upon with disdain because his studies are sometimes not quantitative in method."[11] The eminent physical chemist, G. N. Lewis, has declared, "I have no patience with attempts to identify science with measurement, which is but one of its tools, or with any definition of the scientist that would exclude a Darwin, a Pasteur, or a Kekule."[12] To these three men Cannon would add such other great scientists whose work does not depend on measurement as Harvey, Virchow, Pavlov, and Sir Charles Sherrington. Thus, the use of mathematics is not the only sign of the existence of conceptual schemes and of highly developed science. This is a lesson which some social scientists may also learn, for often their researches seek quantification at all costs, even the cost of scientific relevance. Metrical precision and determinateness are the ideals toward which all science may aspire; they are not, however, the hallmark of useful science.

Since there is some confusion on the point, an even more general statement of the nature of the relations of mathematics and science seems to be required. Mathematics is sometimes called "the only true science." But although mathematics is of the essence of rational and logical thinking, and despite its close connection with science, mathematics is not substantive science at all. It is instead a language, a logic, of the relations among concepts, an extremely useful and precise language which has made possible great advances in many areas of science but which is not to be mistaken for scientific theory. It is true that in physics so much theory is cast in mathematical terms that it sometimes seems to be simply mathematics and nothing more. But there are in addition to the relations among concepts which mathematics expresses so precisely, these substantive concepts themselves: mass, energy, etc. As the modern non-Aristotelian semanticists express it, mathematics is a language of relations, not of classifications and identifications. Mathematics is a language of relations in the same way that Aristotelian and symbolic logic are languages of relations. As such, in sum, it is extremely useful for science, but not to be confused with the conceptual schemes of science. The construction of these conceptual schemes is a difficult enough task in itself, about which we shall say more when we discuss the functions of imaginations in the process of scientific discovery.

The nature and functions of experimentation in science, like those of mathematics, are also sometimes misconceived. It is often thought that experimentation is peculiar to modern science, that it has existed only for the last three hundred years. The extent and elaborate precision of present techniques of experimentation are peculiar to modern science, but the logic and even the practice of experiment are not. All rational thought implies the comparison of like and unlike cases and the consequent assignment of causal priority or functional relationship on the basis of that comparison. In this important sense, experimentation is as old as rational empirical thought and therefore coeval with human existence. The construction of determinately controlled experiments is only the highly developed modern form of what was previously at least implicitly and sometimes explicitly employed. Modern science has been remarkably successful in defining and isolating "concrete" systems of phenomena which correspond precisely to the abstract systems of ideas which compose its conceptual schemes. As Professor H. Levy has indicated, this isolation of systems is highly important for science.[13] Once a system is isolated, by controlled variation of one part, the effect on other parts of the system can be ascertained. In this fashion, experimentation, as we call this process of controlled variation which compares like and unlike cases, discloses the effect of the several variables in the conceptual scheme.[14]

At least this is the situation for much of the physical sciences, but the possibilities of controlled experiments are not nearly so good in the biological sciences. Perhaps this is why the biologist, René Dubos, biographer of Pasteur, is skeptical of the alleged powers of experimentation. "Nor is the experimental method," he says, "the infallible revealer of pure and eternal fact that some, including Pasteur, would have us believe."[15] It needs hardly be said that up to the present time controlled experimentation has been quite rare in the social sciences. These sciences have had recourse chiefly to the logic of comparison among like and unlike cases, with the comparisons unfortunately all too uncontrolled.

The reservation which Dubos has expressed about the experimental method is probably in part a reaction to a certain prevalent exaggeration of its significance in science, especially of its significance in comparison to that of conceptual schemes. It is usually so much easier to see a scientific experiment than the conceptualization which has made it possible that the conceptualization is overlooked. On this view science often becomes a thing of techniques and gadgets, involving a minimum of difficult rational thought. But experiments, as we have now seen, can be scientifically significant only when they are comparing like and unlike cases which are important for the variables defined by some conceptual scheme. Or, to put it in another way, behind every good experiment there is a good theory.

Professor Butterfield has noted this excessive emphasis on the importance of experimentation in the common belief that the essential change accounting for the rise of modern science is the emergence of the experimental method.[16] An instance of this belief is the credence we place in that apocryphal story about Galileo's experiment of dropping weights from the Leaning Tower of Pisa. Actually, behind such an experiment, if indeed it ever occurred, there lay important new ideas, a whole new theory of falling bodies. And so it was with the other important discoveries that occurred in the sixteenth and seventeenth centuries. Great developments in rational thought, especially in mathematics, and new conceptions in astronomy and mechanics, about all of which we shall say some more in our next chapter, provided a new guidance for the continual experimentation which Lord Bacon had recommended in his *Novum Organum*. For centuries, says Butterfield, experimentation "had been an affair of wild and almost pointless fluttering—a thing in many respects irrelevant to the true progress of understanding—sometimes the most capricious and fantastic part of the scientific program."[17] In neither the Mediaeval Period, as we shall see, nor certainly in the Renaissance, did men lack the inventive skill and ingenuity to construct the technical devices for experimentation. "Yet it is not until the seventeenth century that the resort to experiments comes to be tamed and harnessed so to speak, and is brought under direction, like a great machine getting into gear."[18] The dynamic of modern science inheres in the proper interweaving of conceptualization and experiment.

The problem of experimentation in science suggests a related range of problems which arises from the relations between conceptual schemes and what is usually called "technique" in scientific work. Since conceptual schemes have reference to empirical data, there must be observational techniques for gathering these data and other techniques for ordering these data to the appropriate conceptual categories. The relationship between conceptual scheme and technique is not a simple one, however, for although there is a necessary interdependence of the two, there is also a certain degree of independence. Here is another example of the subtle process that is science. Conceptual schemes may sometimes independently, by deductive reasoning, predict data which the available technique either cannot observe or at least has not yet observed. We have recently seen an example of this sort in the award of the 1949 Nobel Prize for Physics to the Japanese scientist, Yukawa. In 1935, entirely by manipulation of the conceptual scheme of physics, which as we have said is now very largely stated in mathematical terms, Yukawa announced the existence of a particle known as the "meson." This sub-atomic particle has since been observationally discovered by its track on photographic plates and it is the center of a great deal of study in contemporary physics. Note that only the existence of the conceptual deduction directed the attention of technique to the verification of this discovery.

Contrariwise, however, available observational and ordering techniques may collect data which cannot be fitted into current conceptual schemes. Indeed, this is happening all the time in science. We shall give many examples of this when we discuss the phenomenon of "serendipity" in science, or the chance occurrence of unexpected discoveries. Sometimes these eccentric data remain unassimilated temporarily, but very often they immediately stimulate a valuable reconstruction of the conceptual scheme. This is one important path of scientific advance. In the late nineteenth century, many scientific observations were made which could not be fitted into the Newtonian conceptual scheme. It was the great virtue of Einstein's theory that it assimilated these previously unexplained observational data.

Thus we see again that science is not simply the collection of a large body of "facts." It is, rather, the collection and ordering of facts in terms of a conceptual scheme, the scheme always being subject to reconstruction as its use or the use of technique result in new facts. Conceptual scheme and techniques are probably never perfectly integrated and it is often from this very discrepancy that fruitful facts emerge. By "fruitful" we mean, of course, facts which serve in the construction of conceptual schemes of ever-greater generality and systematization. In his discussion of what he calls "Certain Principles of the Tactics and Strategy of Science," President Conant has given us a detailed examination of the kinds of relations between conceptual scheme and technique of which we have been speaking.[19] It is a subject which is close to the heart of scientific advance in all fields.

Perhaps we can now see very clearly the sense in which highly developed science based on conceptual schemes of great generality is essentially a dynamic enterprise. The endless making of improved conceptual schemes introduces a dynamic element into the very center of scientific activity. In this way, human rationality takes on the unending power to move heaven and earth, for sooner or later changes in conceptual schemes issue in changes in everyday life and everyday technology. Veblen has said that "the outcome of any serious research can only be to make two questions grow where one question grew before."[20] This is a characteristic of science, this is a dynamic quality it has, that modern man must not only learn about, but learn to live with. For this is the source of the unending social consequences of science. But of this we shall have more to say later, when we discuss these social consequences.

One last problem about the nature of science remains, the problem of the relation between the conceptual schemes of science and the body of belief and knowledge usually called "common sense." We shall see in our next chapter that every human society has, at the very least, a collection of rational empirical knowledge, or relatively undeveloped science. This kind of knowledge, which we may think of as "embryonic" science, out of which more mature science may grow, constitutes a large part of what is usually thought of as common sense and provides a fairly effective guide to action. But though like all knowledge it is based on some implicit, particularized kind of abstraction, the limitations on the effectiveness of such empirical knowledge should now also be clear. For in the degree in which common sense is not generalized and systematic knowledge, as are the conceptual schemes of highly developed science, it is not reliable knowledge, or, as we may now put it somewhat more technically, it is not determinate knowledge. That is, relatively speaking, common sense does not know the determinate conditions under which its assertions of fact and of relationship between facts actually hold. When these unknown conditions change, the facts will change, and common sense, without determinate understanding of what the conditions are, has no satisfactory guide for further action. The unreliability of common sense, its contradictoriness in the face of changes in conditions which it cannot describe, can perhaps best be seen in the large body of inconsistent and contradictory common sense sayings and proverbs. A collection of the innumerable errors, delusions, and misconceptions to which common sense always has been and still is heir may be examined in Professor Bergen Evans' book, *The Natural History of Nonsense.*[21] Seen in this perspective, science is not, as Huxley said it was, simply "organized common sense." That is, science does not have the limitations of common sense. Of course, Huxley's remark holds if one takes it to mean something else that we have asserted, namely that both common sense and science have in part a common origin in human rationality.

We may say that in every society there is *some range* in the determinateness of its empirical knowledge. Where there is only empirical lore, the rational knowledge of common sense will cover only a relatively narrow range of determinateness. But where there is highly developed science, the range is very much larger, all the way from common sense to science itself. This is the situation in our society. Indeed, in our society it has pretty widely become a part of common sense to recognize the greater determinateness which scientific knowledge has. This is of course a happy condition for the further development of science. Common sense may, however, partly oppose science and it is perhaps not impossible that it should turn against it very strongly. This possibility we shall examine later on when we discuss the sources of opposition to science that exist in our society.

Even where common sense accepts the superiority of science, it can have only a vague perception of the grounds for the greater reliability which science has. The conceptual schemes of science are now very highly technical systems of ideas, available only to professionals with long training in the relevant fields. Insofar as the untrained lay person understands these ideas at all, he grasps only some basic idea and not the full technical comprehension. Thus, only the vaguest notions of the meaning of Einstein's theory of relativity can be had on the basis of common sense. Indeed, most of us do not have even a vague understanding of the theory but rather feel about it much the way the men of Newton's time felt about his new and apparently absurd notions that also seemed to contradict common sense. Newton's theory was based on what seemed to his contemporaries to be most improbable assumptions.[22] The notion of a force acting at a distance is quite a different thing from the notion of a direct push, on which our intuitive, common sense understanding of force rests. Gradually, of course, in Newton's case, conceptual scheme and common sense were in some fashion reconciled. "After a generation or so most men managed to convince themselves that action at a distance was a reasonable and comfortable idea."[23] Or at least conceptual scheme and common sense puzzled each other less. Eventually, and this is true for us today, the Newtonian notions came to be regarded as intuitively obvious, as common sense. As Mach has put it, "uncommon incomprehensibility became a common incomprehensibility."[24]

The same process seems to be recurring in the case of the new Einsteinian conceptual scheme. Einstein's assertions about the basic principles of the physical universe conflict with our Newtonian common sense in the same way that Newton's mechanistic views once conflicted with the earlier organismic views of Mediaeval common sense. "There can be very few people, if any, who think naturally or intuitively in terms of a curved universe whose geometrical properties have taken the place of gravitation."[25] Fortunately, although they are otherwise great, the deviations between our common sense notions of space, time, and motion and those introduced by modern physics are negligibly small so far as the experience of everyday life is concerned. One may have a glimpse of how fantastic ordinary life would be, if this were not so, in George Gamow's charming book, *Mr. Tompkins in Wonderland.* Mr. Tompkins' dreams of a world in which the new notions of modern physics are realized could only be dreams and not everyday reality.[26]

Not all new conceptual schemes in science are so revolutionary, of course, as Newton's and Einstein's. Such great scientific syntheses have not occurred very frequently. And yet there is a constant, and perhaps increasing gap between the technical conceptual schemes of science and our everyday common sense. This gap has social consequences for scientist and layman alike, as we shall see later on. But it is, to repeat what we have already said, a fortunate condition for the advancement of science that such a gap is, on the whole, tolerated by common sense. This is so both because of our moral respect for science and because of our conviction that science has great usefulness for the practical problems of human society.

II

The Historical Development of Science: Social Influences on the Evolution of Science

HAVING briefly explored the nature of science—its source in human rationality, the variability in its level of generalization and systematization, its connections with common sense—we have now to make some further discovery of that nature by tracing out the historical process of the evolution of science. This is, of course, a treatment which can only be sketched here. The history of science is much longer and richer than we think, and only in the last forty or fifty years has modern scholarship been writing it for us. Although a very great deal more remains to be done, by now enough of the story has been told so that we can have some sense of how ancient and extensive our heritage of human science is.[1].

Although it must necessarily be all too brief, our account of the historical development of science will stress six major themes which are essential for understanding the social aspects of science. They are all themes or uniformities which have been explicit or implicit in the preceding analysis of the nature of science but which here find historical exemplification. These are the six themes, set down separately, even though all of them, as we shall further indicate, are inter-related:

1. The universality of human rationality.

2. The continuity of the evolution of science.

3. The variability of the levels of activity and accomplishment in science throughout history.

4. The importance of many different social influences on the development of science.

5. The relative autonomy of science considered as one component of society.

6. The reciprocality of influence between science and the other components of society.

Before proceeding to the historical discussion, some general comments on each of these six themes will help to clarify their meaning and their inter-relations.

Perhaps the least needs to be said about our first theme, the universality of human rationality, except as a summary of what we have already said about this subject. It is basic to all that follows to see the source of science in the generic human attribute of empirical rationality. At this point we shall simply note the fact that science has occurred in pre-historic and ancient societies, in so-called "primitive" or non-literate groups in all parts of the world, and in the Graeco-Roman, Mediaeval, and Modern worlds.

The second theme, the continuity of the evolution of science, needs to be stressed because of a provincialism about science which exists in the modern world. Partly out of historical ignorance which could not be avoided until recently, and partly out of a rationalistic bias about the nature of earlier and other societies, many of us have felt that empirical rationality and science are both *uniquely* modern. But in this, as in other respects, there has been no radical discontinuity in history. Not only has some form of science existed in all societies, but the several forms have been built each upon its historical antecedents. For at least the last three or four thousand years, and even beyond that, the record of the evolution of science runs fairly continuously, without unbridgeable gaps. Now running very slowly, now slightly faster, the stream of science may be traced through its constant and cumulative progress. Here, if anywhere, we could profit our understanding by looking at the historical record in its great detail. Our all too slight knowledge of the historical record is too much structured into the delineation of macroscopic "periods" in science: Greek science, Arab science, Modern science; and we do not see how these periods are related and blend into one another. We do not often see how the science of the Ancient Near East was the partial basis of Greek science; how this in turn flowed into Hellenistic science; how the Greek legacy was transmitted by the Hellenistic Alexandrians to the Arabs and thence to the Mediaeval world, which also received the ancient science directly, through the Church; and how, finally, the Mediaeval Church and the Renaissance re-discovery of Greek science made essential contributions to the foundations of modern science. Nor do we see the additions which each period in the evolution made to the whole. The growth of science is more a matter of many small steps than of a few great leaps, it is more like a slowly enlarging coral reef than like a Paricutin, created by explosive volcanic eruption.

Unfortunately, we shall not be able here to trace out the detailed continuity in the growth of science. But if we too must carry on our discussion in terms of large periods and great movements, we shall always try to show the connections between earlier and later events. They are always there, even when we cannot speak of them. And later on, when we describe the social process of discovery and invention, we shall return to this theme of the continuity of the evolution of science, although in an analytical and not primarily historical fashion. There we shall be able to demonstrate the close dependence of each scientific innovator upon those scientists who have gone before him.

To speak so emphatically of the broad unity of the evolution of science is not, however, to deny the occurrence of some diversity in the details of that process. Not every step in the development is made inevitably and immediately upon its predecessor. There have been, in the details of advance, independent lines of development, but in the larger stream these smaller ones all flow into a single great channel. There have been multiple independent discoveries in science, and we shall give a long list of them later, in a more appropriate place in our discussion, but all these are, on the larger view, part of the continuous and unitary evolution of science. Of course as the extent of communication has increased among human societies, the unity of the growth of science has probably also increased. As the many societies of the past have been knit by the ties of communication into the more nearly one world of the present, science has more nearly become a unity in detail as well as in the large. And yet even today, for all the increased effectiveness of communications among societies, the incidental diversity of science remains, witness the persistence of the phenomenon of multiple independent discoveries. As a result of an increase in political barriers to communication, this diversity may even increase somewhat. But this should never obscure the larger perspective, in which science is, by its essential nature, an evolutionary unity.

Nor should our perception of the broad evolutionary unity of science lead us into the error that scientific development is an easy and inevitable thing. C. D. Darlington, the British biologist, has said of this error, "Most people probably imagine that science advances like a steam roller, cracking its problems one by one with even and inexorable force."[2] But a nearer inspection of any single advance in the history of science reveals how false this notion is. Science is always difficult, its evolution is always "halting, complex, almost irrational."[3] How difficult each next step in science is, how much it is not inevitable but requires an act of individual creativity, we shall see when we discuss the social process of discovery later on. And still, on balance, there remains the large evolutionary continuity.

Our third theme, the variability of the levels of activity and accomplishment in science throughout history, is complementary to the one of which we have just been speaking, and the two must always be taken in relation if we are to maintain a proper balance in our view of the evolution of science. Little need be said about this third theme if only because the uniformity it expresses has been typically over-emphasized rather than ignored. It has been over-emphasized to the point where variability has been taken to range all the way from the non-existence of empirical rationality to its high development in modern science. We have now sufficiently indicated that this is a greater range of variability than actually exists. The range of variability which remains is real enough and a constant awareness of this range is valuable if it leads us to inquire into its sources.

And this leads us directly to the fourth theme, the importance of many different social influences on the development of science. Now this is a theme which by this time, and especially when it is put in this very general and rather vague way, may seem like a truism to most of us. Yet this was not always so, not even quite recently. At least as late as the 1920's the view that there was something called "pure science," by which was meant a science entirely uncontaminated by the workings of social factors, this view was widely held. It was perhaps then the predominant view of the nature of science, not least of all among scientists themselves.

Now if our views on this matter have changed greatly, the change cannot be so much the result of a greater intellectual understanding, although this too we do have. The change has probably been even more the consequence of a whole series of social events which have crowded in upon us since 1930. The great world-wide economic depression of the 1930's, with its "frustration of science," of which we shall say more later; the rise of Nazi Germany, with its preachment of an "Aryan science" and its violence toward Jewish scientists; and World War II, with its urgent, large-scale application of science culminating in the explosion of the atomic bomb—all these social events have brought home to scientists and all the rest of us too, in a most immediate way, that there are important social influences on science. The old illusion of a "pure science" is no longer tenable, at least not in the extreme form in which once it was held. In what sense we may still speak of a "pure science" is something which we must postpone for a while. In any case, one has only to read the speeches and writings of scientists since the 1930's—take, for example, the annual presidential addresses of the British or American Associations for the Advancement of Science—to see how the old view has evaporated. Where our social intellect alone had failed us, history has forced upon us this more adequate understanding of the social nature of science.

Yet intellect had not really failed us, for the social view of science had been for some time available in quite explicit form in the Marxian sociology. Marx and Engels themselves had, quite directly and in detail, asserted the dependence of science upon the society in which it existed. This Marxian analysis had been expanded by a group of German social scientists into the study of *Wissenssoziologie,* or the sociology of knowledge, which tried to show how science, as well as other forms of knowledge, were directly affected by social factors.[4] And beginning early in the 1930's, a group of

Marxist-oriented scientists and scholars produced a spate of historical studies which sought to demonstrate what they often referred to as "the social roots of science." Most notable among these studies were those by Benjamin Farrington on Greek science; by the Russian, B. Hessen, on the science of seventeenth century England, especially of the Newtonian physics; by J. G. Crowther on the science of nineteenth century England; by J. D. Bernal on twentieth century England; by Lancelot Hogben on the whole history of science; and, finally, by the American mathematician, Dirk Struik, on United States science in the early nineteenth century.[5] To all these studies, whatever the deficiencies they have, we have an intellectual debt, not only for the specific information they contain, but for what they have contributed to the increase in our awareness of the social connections of science. This positive debt is great even when it is not always obvious. And even the negative debt, that which emerges out of the correction of their errors, out of the refinement of their analyses, even this debt is great because the improvements on their work have served to strengthen the sociology of science in general.

But for a long time the Marxian view of science was rejected, and this for many reasons, not all of them intellectual, of course. Whatever all these reasons may be, and most of them do not concern us here, it is true that one important reason was in fact the intellectual inadequacy of the Marxian view. We have refused to accept the Marxian sociology of science as it stands not wholly because we are irrational or are blinded by the presuppositions of a capitalist society but partly because it does need correction and refinement. We can profit by pointing out certain of these general deficiencies and by stating positively what we can now consider a more satisfactory account of the social relations of science.

The burden of the Marxian view on these matters is that science is a wholly dependent part of society, molded fundamentally by the economic factor; and that therefore there is no reciprocal influence between science and the other components of society. This view is not acceptable as an adequate understanding of these matters. We leave aside the fact that what the Marxian sociology means by "the economic factor" is often an ill-defined category, filled with what are actually diverse elements; for example, sometimes the social relations of production, sometimes technology as such, and sometimes the economic ideals which are prevalent at a given time. It is not correct that only the economic factor, however interpreted, has an effect upon science. As we have stated it in our fourth theme, and as we shall show by our historical description of the evolution of science, many different social factors have had and continue to have an important influence on science. No one of the several alternative factors is necessarily and under all conditions more important than the others. The intellectual, the religious, and the political factors, for example, are no less, and of course no more, influential always than is the economic factor. Now one, now the other, now several of these in conjunction, can be seen to have an effect on the development of science. Indeed, perhaps the hardest tasks for analysis are those in which several factors are working conjointly, often together with an influence from the internal condition of science itself. Yet this remains the proper task of the sociology of science, to seek out the specific conditions under which each one of the several possible social factors or many of them together have actually influenced the course of science.

If we cannot as yet always specify the precise conditions under which the many different social factors do exert their influence, still we can see that they are actually operative in some significant fashion. The political factor, for example, had a beneficial effect on French science in the early nineteenth century. During the Revolution, the Convention set up the *École polytechnique,* in which, for the first time, there was organized the practice of having expert scientists teach students through apprenticeship to actual research. This new practice trained up a generation of excellent scientists and soon this training procedure spread from France to Germany and to England.[6] We shall later on have occasion to refer to many other instances of political influence on science, not least of all in the cases of Nazi German and Communist Russian science. Or take the influence of the intellectual factor on science. We shall see in just a little while how a change in fundamental intellectual assumptions during the seventeenth century, a change to the intellectual conviction that there exists what Whitehead has called "an order of nature," we shall see how this change had a favorable influence on the growth of modern science. Or, finally, take the influence of the religious factor. We have already suggested that the men of the

Mediaeval Period were more interested in religious rationalism than in empirical rationalism. This could not but have a partially and temporarily retardative influence on the evolution of science. And so it goes, continuously,—as we shall see at great length throughout this book,—now one, now another of the social factors has an influence on science, sometimes relatively favoring its growth, sometimes relatively hampering it. This is the inevitable rule, for science is not something apart from society.

We include within our meaning of "the social factors," of course, the economic factor. In our attempt to correct the deficiencies of the Marxian analysis, we must not swing to the opposite extreme and dismiss as unimportant what it so much stresses. For example, as we shall also see again later on, the current state of technology has an important influence on science. The technological possibility of the atomic cyclotron and of the electronic calculating machine in our own day has a most beneficial influence upon the development of physics and the other sciences as well. Or, to take another case from our own time, one which we shall discuss at length later, the support which modern industry gives science has an important effect upon both the rate and the direction of growth in modern science. In these ways and in others, the economic factor is significant for a sociology of science.

We should make it perfectly clear, of course, that when we speak of social influences on science we are not implying anything about the personal motives of individual working scientists. We shall want to say a great deal more about the relation of motives and social organization when we discuss "pure" and "applied" science, but here we need only note that these are two separate questions. "In fact," says Samuel Lilley, an English scientist who has concerned himself with the social aspects of science, "the list of scientists' motives would include virtually the whole range of human desires and aspirations."[7] Social influences operate whatever, and sometimes in despite of, the particular motives of scientists. This is, indeed, only another instance of what is generally true about all kinds of social behavior. And it follows from the existing discrepancies between individual motives and social influences that scientists may or may not be aware that one or another social factor is directly or indirectly affecting their work. "Social movements can influence the work of an individual when he is not consciously concerned with them. In fact the individual's lack of consciousness might in some cases actually increase the social influence, by denying him the opportunity of consciously correcting for its effects."[8] As we shall see in detail later on, the very nature of science makes it inevitable that science will have consequences which were unintended by any particular scientists; by the same process, science is directed and channeled in many ways of which scientists are not aware.

There is one last important characteristic of the social influences on science. This is that the influences are not only diverse but are now stronger, now weaker, never continuously uniform. The degree of social influence on science is a subtle process which we have still only crude techniques to measure; but this present crudity should not blind us to the gradations that exist and are important. Now one social factor may have an important influence on physics, now another; now the relevant social factors are more strongly affecting biology and leaving physics relatively untouched, now they may reverse this relative emphasis, or shift away entirely to some other science, like chemistry. The complexity of the process is great, not in the sense that it is intrinsically beyond our understanding, but in the sense that there are a multitude of different relations between the parts of science on the one hand and the various social factors on the other, and we must recognize as many of them as we can.

And this consideration of the relative strength of social influences introduces us to the fifth of our themes, the relative autonomy of science considered as one component of society. For despite all the social influences that mold its evolution, science always retains a margin of independence, as do the other parts of a society, just because it has an internal structure and a process of action of its own. This internal structure and this special scientific process we shall be investigating all the time in this book, and we shall see how it provides a relative independence for science at the same time that science is interacting with the several other parts of a society. One important element in the relative autonomy which science has is its development of highly generalized conceptual schemes. We may say that the margin of independence which a science has is the greater the more highly developed is its central conceptual scheme, although of course social influences are still operative no matter what the degree of development is. For this reason, the strength of social influences on the development of the social sciences is probably greater now than it is on the physical sciences, because of the weaker conceptual schemes of the social sciences. As they develop, conceptual schemes determine a certain line of development of their own; they do not then shape themselves simply in accord with some "social need." For example, the conceptual schemes of the biological sciences are not yet ready to deal adequately with the phenomena of cancer, although there is a pressing social need for an effective cancer therapy. But there are other important elements in science which secure its relative autonomy besides conceptual schemes. Of these, of such things as the strong values which scientists have and the independent social organizations to carry on their activities, we shall say a great deal more later.

And finally we may state our sixth theme, which has perhaps been implicit all along above because all our themes are significantly interrelated. This is the theme which asserts the reciprocality of influence between science and the other components of society. If science is affected by other social factors, and if it has a relatively autonomous development of its own, it also has an effect upon the rest of society. We have already said that science has social consequences, and this fact will be a recurrent theme here. The view we take is that society, including science now, is a web of interacting structures in which the effects ramify and re-trace themselves time and again. If the lines of influence are hard to trace through their interweavings, it is only because our instruments of analysis are not yet good enough, not because the lines are not in fact interwoven.

These, then, are the six general themes which are important for understanding science as a social activity. They form the groundwork for our whole investigation. We shall first try to show their significance in our sketch of the historical development of science which follows immediately. They will also be our guide-lines in the next chapter, on science in modern society, and still further, throughout the whole book. Perhaps, after the preliminary statement we have given them, the reader will find them apparent in our analysis even when they are not explicitly noted.

Of all the creatures of the earth, man is the only one who is born without elaborate instinctive patterns of adjustment to his physical and social environment. Therefore man has always been, and always had to be *homo sapiens,* man the thinker; and *homo faber,* man the maker. Without the gift of empirical rationality, human life in the face of a resistant environment would be impossible. The gift is universal and aboriginal in man, reaching back to the first moment of pre-history of which we have any knowledge. That is why we can say with Crowther, "Early man's mode of living was impossible without a considerable knowledge of elemental mineralogy, geology, zoology, botany, and astronomy."[9] And, we may fairly add, sociology and psychology, for a certain minimum rational knowledge of the behavior and feelings of fellow human beings is as essential as knowledge of the world of nature.

The two aspects of man's activity, as *homo sapiens* and as *homo faber,* have of course always been connected. Only in modern science do they become somewhat specialized, although they are still, as we shall see, in a close and important connection. Early man was unspecialized, and therefore we must trace his history as *homo sapiens* through his activities as *homo faber.* Only the tools of earliest man are available to us and from these the archeologists have reconstructed the fundamental discoveries made by man in pre-historic times. By Late Paleolithic times, man already had a "vast variety of tools"—axes, knives, saws, spokeshaves, scrapers, mallets, awls, needles of ivory, spears, harpoons, bows, spearthrowers, and even tools for making tools.[10] All through Paleolithic times there was a continual advance in the development of tools for controlling man's environment.

The Paleolithic advance in empirical rationality culminated in what is called the Neolithic or New Stone Age. This was the age that saw the discovery of hoe and digging-stick agriculture. It was made possible, of course, by the invention of special agricultural tools: the hoe, the sickle, the flail, and the quern to grind corn. There was also in this time great progress in the arts of pottery-making, mining, polishing of stone, and spinning and weaving. Indeed, so great are the Neolithic advances in empirical rationality that Lilley refers to this period as "the first great Industrial Revolution." We can get some idea of how slow scientific evolution used to be and how relatively rapid it has been in historic times by noting that this first Industrial Revolution occurred not much more than seven thousand years ago. Conditions for the advance of this Industrial Revolution were especially favorable in the valleys of the Nile and Indus and in Mesopotamia. In these areas there was a "great spate of invention in the couple of millenia before 3000 B.C."[11] In these areas men first discovered how to smelt and use metals, to harness animals, to plow, and to make wheeled carts and sailing ships. As inventions always are, these inventions were all interconnected. For example, metal carpentry tools were needed before plows could be made. Incidentally, the smiths, the workers in metals, who first emerge in this period may be the first occupational specialists in man's history.

Although we have spoken only of their tools, the ages before 3,000 B.C. were also the discoverers of another kind of more general rationality, in the form of mathematics, which was increasingly to be of assistance in the advance of empirical rationality. Developed in close connection with empirical tasks like agriculture and irrigation, mathematics occurs at least as early as the fourth or fifth millenium B.C. among the Egyptians and Babylonians. And, says Struik, who has written of the history of mathematics, "if we assume that mathematics was born when men began to have some understanding of numerical and geometrical relations, then mathematics is much older than those ancient peoples."[12] Its history, he says, may go back even to the Old Stone Age.

In brief summary, then, of early scientific evolution, we may say that on the whole it was continuous but very slow, although there were periods of very much greater accomplishment than others. Empirical rationality pretty much remained particularized, bound up in technology and craft, without reaching high levels of generality and systematization in conceptual schemes. Social influences are hard to specify, but probably the stable social organization and complex division of labor in such societies as those of Egypt and Mesopotamia were especially beneficial to scientific progress. For early societies it is perhaps more obvious than in any other case that the advance of empirical rationality has social consequences. From their earliest developments, science and the rest of society have been in continuous interaction.

We must turn aside for a moment from our account of early science and how it led into Greek science to consider a set of societies which exist at the present time but which are often grouped with the earlier societies of which we have been speaking. These societies are what we may loosely call the "non-literate" societies, which have sometimes been referred to as "our primitive contemporaries." It was the view of the older, social-evolutionary anthropology that these societies were survivals from pre-historic times. Indeed, an extreme statement of this view even had it that contemporary non-literate societies were, like their ancient counterparts, "pre-logical" and irrational. This conception of "primitive" man's thinking has long since been rejected by social anthropology as a result of its study of a large number of non-literate societies in every part of the world.[13] And yet this view lingers on in common sense. So recently as the Princeton Bicentennial Conference of 1947, for example, a distinguished natural scientist spoke of the progress we have made "from the mental attitude of the savage, where demons lurk behind every bush." Actually, for all his magic, non-literate man possesses a great deal of rational empirical knowledge.

For example, a systematic survey of invention in non-literate societies adduces a wealth of corroborative evidence for the conclusion that "invention is indigenous in the nature of man."[14] Here are some of the areas in which this survey shows that non-literate man has a considerable body of rational technology: tools and mechanical devices, the uses of fire, stone-working, the potter's art, the uses of plants, the making of textiles, the capture and domestication of animals, and devices for travel and transportation.

Or we may take the evidence from a single field of rational activity, the field of medical therapy. Another anthropological survey shows how extensive is non-literate man's rational equipment in this field.[15] Including all his societies, primitive man has discovered the following drugs as specifics in therapy: quinine, curare, opium, and digitalis. So serious a surgical operation as trephining of the skull has been practiced in the earliest times. Among the Ashanti of West Africa, inoculation for snakebite is successfully performed. Such techniques as cupping, bloodletting, suturing and stitching of wounds, cauterization, and bonesetting are widely used. Medicines are given in the form of decoctions, poultices, embrocations, salves and infusions. Non-literate man knows about hydrotherapy, dietetics, and massage. And, finally, we find him using fumigations, inhalations, snuffs, and nasal douchings. The total of knowledge is indeed impressive.

Or, for a last example, we may consider a single society, the Eskimo. Considering their inventions—windows without glass, the carpenter's brace, the first decked boat, a type of self-supporting vault unknown to civilized architecture, drilling a curved hole—the anthropologist, Kroeber, remarks: "It is not amiss to say that they have produced more inventive geniuses, man for man, than any other people, not excluding the Anglo-Saxon race."[16]

Bronislaw Malinowski's summary statement on this problem of non-literate man's rationality is still the classic one. He speaks from his experience among the Trobriand Islanders of the South Pacific, but his remarks have a general reference. "If by science be understood a body of rules and conceptions, based on experience and derived from it by logical inference, embodied in material achievements and in a fixed form of tradition, . . . then there is no doubt that even the lowest savage communities have the beginnings of science, however rudimentary."[17] In the Trobriand Islands the native shipwright shows his knowledge of the principles of buoyancy, leverage, and equilibrium in his construction of the outrigger canoe. In a crude and simple manner, using pieces of wood, his hands, and a limited technical vocabulary, the shipwright explains some general laws of hydrodynamics to his helpers and apprentices. This science is not, continues Malinowski, "detached from the craft, that is certainly true, it is only a means to an end, it is crude, rudimentary and inchoate, but with all that it is the matrix from which the higher developments have sprung."[18]

Now all this does not mean, it should be noted, that there was not a great deal of magic in earlier and in non-literate societies. But it should be clear now that the existence of magic is no evidence for the absence of rational empirical knowledge. Despite the disapproval of magic in our own highly rational society, some of our health and love practices still have magical elements. Magic is not the product of a mind wholly incapable of empirical rationality. Insofar as he has it, non-literate man uses all the rational knowledge he has for his empirical ends. *In addition,* where he lacks fully adequate rational knowledge for such ends or where a large degree of uncertainty of success still remains in some important empirical enterprise, say in the planting of food crops, in these circumstances he uses magic. Though its scope is larger than in our society, partly because we have more science to achieve empirical ends, magic in non-literate society is clearly distinguished from rational knowledge. Magic has its social functions, as much as science has. Both are necessary for successful social behavior, and this is only especially true where science remains relatively undeveloped.[19]

We may now come back to our account of the main line of scientific development. We come back to the Greeks, who were the inheritors of the great scientific legacy of their ancient predecessors, a legacy which had been greatly enriched by the discovery of iron, a metal which first came into widespread use after 1100 B.C. To this heritage the Greeks added their own original contribution. Whereas before the age of the Greeks rational empirical knowledge, however extensive, had been essentially particularized and specific, a thing of tools and lore, now there is an important change in scientific evolution. For the first time in human history, we find in Greek society the large-scale development of general and systematic formulations of rational knowledge, empirical and otherwise, and this for their own sake. They were the first people to desire, says Taylor, in his history of science, "to make a mental model of the whole working of the universe."[20].

With the Greeks, also for the first time, we come to a period in the evolution of science when there is so very much scientific achievement and so much historical evidence about it that the record becomes somewhat confused. Historians of science have begun to work on this period, but their product is still unsatisfactory; it tells us more adequately what happened than how it happened. Especially in the matter of the social influences on science we still wait for a satisfactory analysis of the Greek accomplishment. Despite these shortcomings, certain important general facts seem to be clearly established. For one, the Greeks made enormous advances in the development of philosophy, logic, and mathematics—those forms of rational thought which are fundamental auxiliaries in the construction of empirical science. For another, logic and mathematics apart, the Greeks definitely made many important discoveries in empirical science, discoveries without which the whole course of scientific evolution would have been much slowed and changed. And, for a third important fact, Greek progress in science was pretty continuous over a long period of time, and always at a relatively high level, even though there was some fluctuation in that level. Let us consider each of these general facts about Greek science, and a few others, in some detail.[21]

Any history of Greek thought probably needs to say least of all about the great new heights reached in logic, in philosophy, and in mathematics, for this is what is usually emphasized in Greek history and therefore it is what we all find most familiar. If our knowledge of Classical Antiquity is not so widespread and ready as once it was, still we all are somehow acquainted with the magnificent accomplishments running almost a thousand years, from the sixth century B.C. to the fourth century A.D., the accomplishments of Thales and Heraclitus, Pythagoras and Parmenides, Democritus, Socrates and Plato and Aristotle, and Euclid and Archimedes. The Euclidean geometry, deriving the whole of geometry by logical deduction from a small number of definitions, postulates, and common notions, may be taken as a type case of the power of Greek rational thought. Another remarkable case is the Democritean atomic theory, a most elaborate speculative theory of the structure and process of the whole universe. There are, however, a dozen other cases that at least equal these two in brilliance and scope in the works of the other men we have mentioned. Had there been nothing more than this great development, Western civilization would still be greatly in debt—as indeed it actually is for this wonderful legacy—to the new power and techniques of rational thought created by Greek society.

But there was more than this. There were advances also in empirical science, advances that we usually overlook, because we are dazzled by the Greek success in rational speculation and because we compare Greek science not with what went before but only with the still greater achievements of modern times. As early as the fifth century B.C. there is the highly rational therapy of Hippocratic medicine, solidly based on generalized knowledge of biology and physiology. Some one hundred years later, and surely based on Hippocrates, there is what Farrington, the historian of Greek science, calls Aristotle's "tremendous achievement in the field of the biological sciences."[22] Aristotle left his school a tradition of organized research; part of the equipment of his school were a library and laboratories. Indeed, this line of development in the biological sciences and in medicine continues to develop throughout our whole period, reaching another peak in the work of Galen in the second century A.D. Empirical science flourished also in other fields. Lilley points out that in the three centuries after Aristotle more inventions were produced than in any comparable period between 3,000 B.C. and the later Middle Ages.[23] Of these we shall speak again below. In the two hundred years after Aristotle's death in 322 B.C., the Lyceum, which he had founded, and its successor, the Museum of Alexandria, produced "a succession of great organized treatises on various branches of science—botany, physics, anatomy, physiology, . . . astronomy, geography, mechanics" which constitute what Farrington designates "the high water mark of the achievement of antiquity and the starting-point of the science of the modern world."[24]

Some may grant the importance of these advances for empirical science but still ask, did Greek science use experiment? Science is nothing, they hold, if not experimental. In this respect, too, what the Greeks did is definitely science. Hippocrates and the other doctors used experiment in the sense of comparison of like and unlike cases all the time. Empedocles' use of the *klepsydra,* or water clock, to establish the corporeal nature of air, is an example of a type of experiment which is more familiar to us because we think of experiment as necessarily using tools and instruments. "With the name of Strato," who was the successor of Aristotle in the Lyceum, says Farrington, "we reach the point at which Greek science fully establishes a technique of experiment."[25] Archimedes also was devoted to experiment. Now all this is not experiment as a certain modern view has it, based on highly generalized conceptual schemes, always highly controlled, and using elaborate physical instrumentation. Yet in its essential logical nature, as providing a basis for scientific inference, the Greeks certainly knew experimentation.[26]

Perhaps Greek empirical science has been underrated also because it did not produce what modern science has given us in such great abundance—the basic need of the machine, prime movers. Even windmills were unknown, despite the use of sailing vessels, although the water wheel was invented about 100 B.C. But the Greeks did develop many other instruments and a few labor-saving devices, perhaps the most notable of which is the Archimedean screw-pump. There was also the screw press, war-engines (catapults and siege-instruments worked by compressed air), graduated water clocks, balances, compound pulleys, the ruled straightedge, and various angle-measuring instruments. The wonderful work in astronomy, most highly developed in the Ptolemaic theories of the second century A.D., however, did not have the advantage of the telescope. But certainly all the new tools and instruments did have considerable effect on everyday life, although not of course the great impact which came about in modern times with the discovery of prime movers like the steam and internal combustion engines. For all its progress, Greek science was not yet ready for this.

We have already said that the development of Greek science covered nearly a thousand years' of continuous evolution. As always in science, there were periods of greater and less activity and advance during this time. The peaks, perhaps, are such bursts of effort as the School of Miletus (sixth century B.C.), the Athenian Academy of the fifth and fourth centuries of the same era, the Lyceum which succeeded the Academy, and the Museum of Alexandria, where what is often called "Hellenistic science" flourished. Indeed, it was probably in the Museum, that occurred what was the greatest amount of scientific activity the world had ever seen in one place up to that time. Its libraries had half a million papyrus rolls and they were used by about one hundred professors whose salaries were provided by the King. There were special rooms for research, anatomical demonstrations, lectures, and study. There were an observatory, a zoo, and a botanical garden attached to the Museum. "Such opportunities for research and scholarship had never existed before. Good use was made of them."[27] We shall see that it is one of the special advantages of modern science that such opportunities and such facilities for research are extremely widespread.

Having described Greek progress in both rational thought and in empirical science, what summary characterization can we give of Greek science? The predominant and most widely-held view is that there was a lack of thorough-going concern in Greek thought to test its generalizations empirically. The Greeks, it is usually held, were more interested in the inner consistency of a system than in objective experiment; they appealed primarily to "some subjective sense of fitness," and to "the argument *ex consensu gentium*."[28] As Whitehead has put it, briefly and perhaps too strongly, the Greeks were "over-theoretical." On balance, this is probably a just description. However, it should never cause us to think that the Greeks were "just philosophers," and not excellent scientists as well. Lest we do, we should note Farrington's summary: "In the Lyceum and the Museum the prosecution of research had reached a high degree of efficiency. The capacity to organize knowledge logically was great. The range of positive information was impressive, the rate of its acquisition was more impressive still. The theory of experiment had been grasped."[29]

On one important point we cannot follow Farrington, however, and that is in his analysis of the social influences on Greek science. According to Farrington, and his explanation has been adopted by others, Greek science declined when it became a society divided into freeman and slave. This is said to have happened in Plato's time, and upon Plato is laid the heavy burden of blame for constructing an ideology justifying the superiority of citizen to slave, of theory over practice, of philosophy over science. Yet Farrington's own record of Greek history belies the claim that Greek science declined after Plato. For example, he praises the men of the Alexandrian Museum, he praises Ptolemy and Galen for their scientific achievements, their observations and experimentations, yet they lived some four hundred years after Plato. The interpretation of all of Greek science as the product of a certain kind of class structure and its accompanying ideology seems to be excessively simplified by a doctrinaire Marxism in Farrington's point of view.[30] The existence of a slave society may have been significant, although there was continuous and large scientific development all during the period that this kind of society prevailed. A satisfactory sociology of Greek science still remains to be written. When it is written, not only will it include more social influences than the class structure in its account but also it will have a vital sense of the way in which the relative autonomy of science interacted with these social influences to produce the Greek science we have been examining.

The historical successors of the Greeks were, of course, the Romans, who were chiefly distinguished for their achievements in law, administration, and the military arts, and of whom it is well known that they made no advances in philosophy or science, in mathematics or technology. The Romans had other gifts to make to Western civilization than the Greeks; but though they did not themselves become scientists, they at least supported the science already in existence. It is no small virtue of the Roman Empire that it permitted Hellenistic science to flourish during the first three or four centuries of the Christian era. After the Romans, of course, in the so-called Dark Ages that lasted for perhaps five centuries after 500 A.D., there was a considerable decline in scientific activity. Yet even this period of apparent decline needs to be seen in a broader perspective than the one conventionally offered us by our history textbooks covering the history of Europe from the fall of Rome. For if we include the whole Mediterranean world in our view—and why should we not?—and not just the continent of Europe, then we must take note of a fairly high level of scientific activity in this period by the Arabs, who spread their new religion all along the southern rim of the Mediterranean and eventually penetrated Europe through Spain. Where they were an offshoot of Christianity in religion, in science the Arabs adopted the heritage of Hellenistic science which the successors of the Romans, in comparison, greatly neglected. And the Arabs made important contributions to the evolution of science. Not only did they make progress in medicine, in biology, and in all the technical arts, but they discovered algebra and they invented the zero, thereby giving to the world the decimal system in mathematics, the system which was to make scientific progress henceforth infinitely easier than it had been before.

In short, when we consider Arab science—a much-neglected subject which deserves more extended treatment than the few words we have given it here—the historical evolution of science is much less discontinuous than it sometimes appears. When we consider the Arabs as part of Western history, we see that the Dark Ages were darker in Europe than elsewhere. We are not then unprepared for the essential contribution which the Arabs make not only to Mediaeval but to modern science. A history of science without the Arabs taken in relation to the rest of what occurred in this period is a poor history indeed.

By the Middle Ages the main stream of scientific evolution has, however, come back to Western Europe, where it has remained ever since. This may seem strange to some, that we find this early source for scientific advance in the Mediaeval Period. It may seem strange because the notion is so common that this was a time that was at best un-scientific and at worst anti-scientific. Yet if we take the broader view of science that we have here adopted, the view of science as *one* form of rational thought which is continually enriched by progress in the other forms of rational thought, then we can easily see the enormous contribution that the Middle Ages has made to the development of modern science. To the Middle Ages we owe not only a great enhancement of our powers of rational thought, so useful later on in empirical science, but also the fundamental conviction that these powers are inalienable capacities of man in society. These are gifts we often too cheaply value. And, we shall also see, there was, even, more empirical science and technology in this period than our school histories customarily report.

The interest in and the development of empirical science, we should now readily accept, is always a matter of degree, not of absolutes. On this view, it is true that Mediaeval Western society was much more—but not completely, as is sometimes thought—interested in the supernatural than the natural world. And therefore it was in the area of the religious and the supernatural that the Mediaeval Period, building upon its heritage from the Greeks, and especially from its master, Aristotle, developed the power of rational thought to such a high level. Indeed, perhaps because of our empiricist bias, Mediaeval Scholastic philosophy has become a kind of symbol for the nature of fine and excessive rational speculation. There is probably no greater single achievement in the history of rational thought than the monumental system of writings of St. Thomas Aquinas. And this was only the greatest among a large number of wonderful accomplishments by the Scholastics.

Now how is this advance related to the evolution of science which we are here tracing? In that book of his which has had so great an influence on modern thought, *Science and the Modern World,* Whitehead has shown the significance of Mediaeval thought for modern science.[31] Something more is wanted for science, says Whitehead, than a general sense of the order in things. "The habit of definite exact thought," so essential in science, "was implanted in the European mind by the long dominance of scholastic divinity." And, fortunately, the habit remained long after the philosophy had been repudiated, "the priceless habit of looking for an exact point and sticking to it when found." Whitehead meantions ofter gifts of Mediaeval thought to the development of science. For example, "the inexpugnable belief that every detailed occurrence can be correlated with its antecedents in a perfectly definite manner exemplifying general principles," while existing to some extent in all societies, was broadened in its application by Mediaeval society to include all of nature. A fundamental conviction of Mediaeval thought, now transmitted to science and modern thought generally, was that "there is a secret, a secret which can be unveiled." This is not universally and in all societies an equally powerful habit of mind. The Mediaeval thinkers who insisted on the rationality of God and on the reflection of this rationality in Nature are, for example, different from the great Oriental thinkers who have seen only an inscrutable force in Nature. The implications of these two radically different conceptions for the development of empirical science have been made clear by history.

Now to say all this, as Whitehead has, is not to say that the Mediaeval thinkers themselves favored empirical science and it is certainly not to say that they did not consider religious problems more important. The favorable influence of Mediaeval religious conceptions on Western science was unintended, though large. "The faith in the possibility of science," says Whitehead, "generated antecedently to the development of modern scientific theory, is an unconscious derivative from mediaeval theology." Sometimes the most powerful elements in society are those which are unintended. This is often the case with those basic cultural values about rationality and Nature of which we are "unconscious" because we take them so much for granted. In our next chapter we shall consider explicitly the set of basic cultural values which have made science so congenial an activity to the modern world.

We have said that interest in science is a matter of degree and that it is not true that the Middle Ages were absolutely uninterested in science. In the midst of the predominant concern they had for rational understanding of the religious and the non-empirical, these times saw a slow growth also in rational knowledge and control of the empirical world.[32] Our superficial knowledge of the Mediaeval world has made us exaggerate its rigidity, its lack of change and progress. This is a distorted picture which new work in the history of science is changing. In the lay world outside the meeting-places of the Scholastic philosophers, and even in the religious monasteries themselves, subject as they were to the rational discipline of St. Benedict, numerous advances in generalized empirical knowledge and their related improvements in technology occurred. The "extraordinary list" of inventions which were made in this period has long been overlooked. Between the ninth and the fifteenth centuries of our era there were invented: the modern method of harnessing an animal for riding with saddle, stirrups, bit, and nailed shoes; also the modern method of harnessing draft animals, with shoulder-collar, shaft, and disposition in file; the water-mill and the wind mill; the mechanical saw; the forge with tilt-hammer; the bellows with valved sides; the pointed arch and window glass; the domestic chimney; the candle and taper; paved roads, as distinct from the Roman walls of concrete buried in the earth; the wheelbarrow; spectacles; the wheeled-plough; the rudder attached to the stern-post of the ship, not the ancient method of steering by oar; the canal lock; powder; the plane; the brace and bit; the lathe, and nuts and bolts; and, perhaps most important of all, printing from movable type.

This is indeed a remarkable list, so impressive in fact that Lilley has spoken of it as the beginning of The Second Industrial Revolution.[33] There was a great shortage of labor in the early Middle Ages, and this, together with all the technical progress, resulted in a very much greater use of water-power, wind-power, and animal-power. For example, as early as 1086, there were 5,000 water-mills in England alone, used for the fulling of cloth, for trip- and forge-hammers and for pumping and winding. The new technical developments and especially the new sources of power greatly contributed to the development of early machines and thereby to the rise of modern science.

Thus, in sum, considering the great growth of rational thought in the Mediaeval Period and considering its advances also in empirical science and technology, we see again that there has been no radical discontinuity in the evolution of science. From its earliest history, Western society and its antecedents have experienced continual, if sometimes slow, progress in rational empirical thought and in the control it gives over Nature. Each age has made its contribution to the stream of development; and in modern times the consequence is a broad river of new knowledge and new applications of that knowledge.

In our sketch of the historical development of science we come now finally to the period roughly included in the sixteenth and seventeenth centuries, the period which is usually inclusively labelled as "the rise of modern science," and here the inadequacies of a sketch become nearly overwhelming. Here certainly there is a need for at least a whole volume, a tight-packed volume. And yet, for our present purposes, certain important points need to be understood, the significance of some of our themes needs to be revealed in the historical record. But we cannot too much emphasize that this should only be the prolegomenon, should only be the invitation, to the further study of the history of science in these two wonderful centuries.

We need, first of all, to understand the relation of the many events that make up the rise of modern science to all that has gone before. It should of course be clear by now that this was not something that occurred "ex nihilo," a wholly strange or new phenomenon in human society. More specifically, a closer reading of the history of science than we usually make will show that we have marked off this period too sharply, that the Middle Ages and the early modern period run into one another in a great many ways, and not least of all in scientific development. This is to say, once again, that the evolution of science has not been discontinuous. And yet "something big" did occur in the sixteenth and seventeenth centuries, something so large that it seems to some to be a "mutation" in the evolution of science.[34] The basic historical fact is as striking as anything can be, the fact that one of the most important changes in all history occurs at this time. "Since the rise of Christianity," says the English historian, Butterfield, in his recent book, *The Origins of Modern Science,* "there is no landmark in history worthy to be compared with this."[35] Science takes on a new scope and a new power both so great that the evolutionary change in quantity seems to be almost a change in quality. But we may accept the metaphor of "a mutational change" only with caution. We may accept it, that is, only if we think of a mutation as related in a fundamental way to its antecedents, and not otherwise. For then we should do violence to the continuity of scientific evolution and to the absence of any radically new occurrences in the realm of empirical rationality.

We must adopt another caution too. We may profitably think of the rise of modern science as a "mutation" only if we understand that it is related to its concomitants, as well as to its antecedents. Here we are recalling our theme that science, while in part independently developing through its own structure and logic, is also constantly interacting with a great many concomitant social factors. We have already noted that it is easy to think of science as a whole in too simple a fashion. So also is it easy to think too simply of what the rise of modern science means. This great complex of events does not limit itself to a single man, like Newton, or even a small group of men, including Kepler and Boyle; it is not limited to a single branch of science, like physics; it does not occur in a single country, like England or France; and it certainly is not entirely explained by some single, or even a few, social or economic or religious changes. The rise of modern science, even in its most narrow delimitation, covers two centuries. It involves a multitude of social changes and a multitude of scientific changes, many of them proceeding on their own course, but many also interacting continuously.

We have chosen to stress the complexity of what is represented in the simple phrase, the rise of modern science, only because the opposite view is the common and, we believe, the misleading one. A notable and valuable exception is the book by Professor Butterfield, to which we have just referred. "The historical process," says Professor Butterfield, "is very complex. While the scientific movement was taking place, other changes were occurring in society—other factors were ready to combine with it to create what we call the modern world."[36] And he has also noted the interaction of these several factors, the scientific and the others. "Indeed, the scientific, the industrial and the agrarian revolutions form such a system of complex and interrelated changes, that in the lack of a microscopic examination we have to heap them all together as aspects of a general movement."[37]

The perception of an existing complexity is often the beginning of understanding: it can indicate the real nature of the problems that confront historical and sociological research. We can do a little better—as does Professor Butterfield himself—than take the several parts of the rise of modern science and "heap them all together as aspects of a general movement." We can make a first approximation to the isolation of *some* of the significant factors in the complex process. This is by no means a full or adequate account of what occurred in the sixteenth and seventeenth centuries; it only shows the direction which such an account would take. We may conveniently group the factors we shall mention—which are, it should be noted, not all of equal importance necessarily—into two rough categories: the internal and the external factors. The internal factors include those changes which occurred within science and rational thought generally; the external include a variety of social factors. The two kinds of factors are separated only for analysis, of course; over the course of the period we are considering they often interacted with one another to produce the final result of modern science.

First, then, we may take some of the internal factors, those which have to do with the relative autonomy of science and rational thought in general. Here one of the fundamental changes which occurred was the emergence of the Cartesian philosophy, a new philosophy for science and rationalism. Descartes did have great debts to the older Scholastic philosophy. Indeed, he had been trained by the Jesuits, and St. Thomas' *Summa* was one of the few books that he carried with him. But his philosophy made a sharp break with Scholasticism by rejecting final causes and by stressing the obligation to seek out the necessary connections among events through precise observation and rigorous logical and mathematical calculation. His conception of the reign of mathematical law was opposed to the older historical traditionalism and became an important guide for the new science.[38] In his great concern for mathematics, of course, Descartes struck a note which was most harmonious for his time. The sixteenth and seventeenth centuries were a period of important discoveries in mathematics, not the least of which was the invention of the differential calculus. This discovery, incidentally, was made independently by both Leibniz and Newton. The differential calculus was an indispensable tool of the new substantive theories in science, especially in physics and mechanics. Astronomy particularly was aided by the new developments throughout mathematics. Other sciences, however, were not yet ready to make much use of such sharp tools of analysis. "Without the achievements of the mathematicians," says Butterfield, "the scientific revolution, as we know it, would have been impossible."[39]

It is of the new theories in science that we may now speak, the new conceptual schemes that achieved an order of generality and systematization that had never before been reached in empirical science. The rise of modern science consists in part in this efflorescence of original conceptions that has been equalled perhaps only in our own time by the developments in relativity and atomic theories. It was an "age of genius," and the accomplishments of giants like Copernicus, Kepler, Galileo, and Boyle followed each other in dazzling succession until they culminated in the magnificence of the Newtonian synthesis, a fundamental groundwork for science which was to go unchanged for over 200 years. All these new conceptual schemes built partly on what had gone before, yet they were also the product of imaginative creativity by individual geniuses. Discoveries in science never simply "have to happen." But of this problem, of the relation between individual creativity and scientific inevitability, we shall say more later. Now we need only note the triumphs of scientific theory, the triumphs not only in individual theories but also, and perhaps even more important, the triumph of theory in general. Although it has its anti-theoretical, empiricistic biases as well, modern science is essentially marked by its understanding of the primary significance of theory for all research.

Modern science is also characterized by refined experimental techniques. In this area too, progress was made in the sixteenth and seventeenth centuries. There was a new importance attached to systematic controlled experiment and there was a new generalized understanding of the method of experiment. "The secrets of nature," said Francis Bacon, the chief exponent of the inductive, experiment method, "betray themselves more readily when tormented by art than when left to their own course." In the new societies of scientific amateurs which grew up in this period, in the newly-founded Royal Society, for example, Bacon's "new philosophy," as it was called, was taken quite self-consciously as a fundamental canon of research. In the work of such men as Boyle and Robert Hooke and Huygens the results were excellent. The new experimentation was, of course, considerably enhanced by the discovery of wonderful new instruments of observation and measurement: the telescope and miscroscope, the thermometer and barometer, the pendulum clock, and the air pump. Here we see, too, how technology and science influence one another and make each other more fruitful, for these new instruments were often partly the result of technological changes. For example, the improvements in the Dutch glass-making industry in the sixteenth century made the telescope and microscope possible. And the navigational requirements of the expanding sea commerce of the time helped stimulate the invention of the pendulum clock. But we must not think that scientific instruments are only the product of craft and industrial technology. The barometer, for example, like many scientific instruments today, arises out of the internal needs and creativity of scientific research itself.

We may see, then, in what consisted the importance of the internal changes in rational thought and empirical science during the sixteenth and seventeenth centuries. It consisted in making explicit the virtues of *coupling* rational thought with direct observation of the empirical world. This was a new emphasis, and the men of the time, realizing the novelty of this powerful combination, might well speak of the "new philosophy." Whitehead has best of all characterized "this new tinge to modern minds." It is, he says, "'a vehement and passionate interest in the relation of general principles to irreducible and stubborn facts. It is this union of passionate interest in the detailed facts with equal devotion to abstract generalization which forms the novelty in our present society."[40]

Running parallel and intermixing itself with these internal changes in science and rational thought was a whole series of important external changes. Even without science, perhaps, the sixteenth and seventeenth centuries mark a great turning-point in Western history. Of the very many new developments in the other parts of society which were provocative of or congenial to the rise of modern science, we may, however, single out just a few. The whole story of science and society in these times has not yet been written.

One important change was an aspect of the Renaissance, the great revival of interest in the ancient works of knowledge and thought. The Renaissance was interested in these works for their own sakes, not in Mediaeval commentaries on them, and it read them with a fresh, critical spirit which made the old knowledge more available for modern use. This was as true for science as for other fields of thought. The works of Archimedes, for example, were translated in 1543 and thus entered more directly into the stream of scientific evolution than they had for almost a thousand years. The new translations, moreover, were printed, and printed usually in the vernacular languages, and so became accessible to new companies of men, men challenging the old authorities or putting rational knowledge to new uses. The Renaissance, then, especially in Italy but also elsewhere in the emerging national societies, had a freshening influence on science as well as on art and literature.

One focus of many of the social influences on the rise of modern science may be found in the new societies of amateurs devoted to cultivating the "new philosophy."[41] The amateurs were found in all countries, Italy, France, England, Holland and Germany; and everywhere they organized societies in which they joined themselves in scientific enterprise and experimentation. In Italy there was the *Accademia del Cimento;* in England, the Royal Society, which is still in existence and the proud possessor of a noble tradition; in France, the *Académie des Sciences,* which flourished only during the lifetime of its spiritual father, Colbert; and in Germany, the *Academia Naturae Curiosorum.* There were many similar, smaller, more short-lived societies in all these countries and in others as well.

These societies provided the first beginnings of scientific specialization. Science was becoming large enough and technical enough so that it was full-time work. Hence the initiation of that trend toward professional specialization in science which is so essential a characteristic of modern science and of which we shall say a great deal more later in speaking of the social organization of science. The societies also became the channels not only of intra-national but of international communication in the new knowledge. Each society had regular foreign correspondents charged with reporting events in his country; and reading the letters of these correspondents was a feature of the meetings. For example, in the late 17th century, every important scientific experiment and article was reported in this fashion to the Royal Society just as soon as it occurred on the Continent. When scientists travelled, they found that they were known and studied in other countries, and they were invited as honored guests to describe their scientific work.[42] The societies published the first scientific journals, of which one may still read the *Philosophical Transactions* of the Royal Society, and they published scientific books by their own members and by foreigners alike. We may recall that it was at the urging of the Royal Society that Newton first published his new discoveries, which he had made many years before.

Ornstein tells us that "it cannot be sufficiently emphasized that it was the experimental feature of science which called forth the societies." This was certainly an important part of their nature, for with the growth of experimental science, scientific laboratories and instruments were both essential and costly. Only joint groups could finance suitable work-places and necessary instruments like air pumps, telescopes, and microscopes. But the societies were more than experimental; they were anti-authoritarian in general. The societies were, for example, a safer place to challenge the old authorities in thought than were the contemporary universities, where Aristotelianism and scholasticism still prevailed. The new science comes chiefly from the laymen of the societies and not from the established universities, as science in modern times does. Still, the universities were not entirely without beneficial influence. Copernicus and Galileo were at the University of Padua at important periods of their lives, and of course the medical school of that university has the glory of having nourished Vesalius, Fabricius, and William Harvey, the latter the discoverer of the circulation of the blood. This conception was a radical challenge to the ancient authority of Galen and Aristotle. It expressed the "radical doubt" preached by Descartes fortified with the confirmation of experiment and observation.

For all their devotion to science for its own sake, the amateur societies were a manifestation also of the utilitarian bent, that other important value of modern science. "They concerned themselves," says Ornstein, "about matters of homely interest such as trade, commerce, tools, and machinery, and tried to improve everyday life by the light of science." Pure and applied science, about the relations of which we shall say more later, had their important connections in those days as much as in ours. This point has, as we have said, been overdone by certain Marxist writers, notably the Russian, B. Hessen. The English historian, G. N. Clark, in rebuttal of Hessen, has admitted the connections between science and technology in the seventeenth century, but he thinks they were not of a general and mutual kind but rather more "piecemeal" than they are now.[43] Especially close were the connections, in England at least, between science and the technology of navigation. The rising English maritime interests—of both her commercial and navy shipping—required a more reliable method of navigation than could be had without good chronometers and easy techniques for determining longitude. These requirements were a direct stimulus to scientific work in this area and so we owe to them advances in the basic science of astronomy and fundamental discoveries on the nature of the spring, the latter making possible finally the construction of accurate chronometers.

We may take another example of the influence of economic and technological factors on the rise of modern science. Lilley has pointed out that from roughly 1550 on, men in many countries were trying to develop new sources of power for the heavier and more powerful machinery which was coming into use.[44] In the expanding mining industry, for example, because of the great increase in the use of coal and the greater depth of the mines, there was a need for a more efficient pump than the old Archimedean screw type. There was also a need for a more efficient form of power than human beings and animals. No satisfactory solution for the latter need was to be achieved until the late eighteenth century, but the seventeenth century did see early forms of the steam engine like the Newcomen and Savery machines. The need for a pump, however, as we have already said, was more speedily met. In the sixteenth century, Archimedean pumps were found everywhere and became very familiar tools. Galileo seems to have learned from the pumps he saw in operation that no suction pump could raise water more than thirty feet. He tried to explain this, but he had the wrong theory. His pupils, however, Torricelli and Viviani did construct the proper theory of the relation between water level and atmospheric pressure. This theory not only made possible the construction of a suction pump for industrial use but had fruitful consequences for scientific theory itself. For on the basis of this theory, Torricelli constructed the barometer, a valuable tool of early research. And von Guericke used the now well known pump in the seventeenth century to make a vacuum. This led to the development of the so-called "'air pump," which later, Lilley says, "in the hands of men like Boyle became perhaps the most important instrument in the advance of 17th century science."[45] Science and technology are mutually helpful.

All this is perhaps enough to say about the economic influences on sixteenth- and seventeenth-century science. The economic changes of those times were very great, and they could not help but affect the newly rising science directly and indirectly. Mercantile capitalism and exploration are very obvious background factors which need to be considered when we are talking about any kind of change in this period. But it will not do to explain any and all changes simply through those factors. These and other external elements were interwoven with internal developments in science itself.

One last external influence on the rise of science should be mentioned, the religious factor, and here we are fortunate in having available some sociological investigations which are models of what the attempt to relate science and social factors should be. These are the studies by Max Weber, the German sociologist, in the comparative sociology of religion and the study by Robert K. Merton, which follows up Weber's lead, on *Science, Technology and Society in Seventeenth Century England.*[46]

As a result of his extensive historical researches on the "great religions" of the world—Hinduism, Confucianism, Christianity, Judaism, Islam—Weber had concluded that the religious values and attitudes of different societies, especially the different views they took on the meaning of Nature and its relation to the supernatural, had a great significance for everyday activities. This may seem a fairly obvious notion to us nowadays, but in the late nineteenth and early twentieth centuries, when Marxist and English Utilitarian philosophies about the primary importance of economics prevailed, Weber's theories were not so readily accepted. Weber was particularly interested in the effect religious values had on everyday economic activities, and he argued that an essential stimulus to early modern capitalism was the new complex of religious attitudes which emerged in the sixteenth century in Calvinist Protestantism. This is Weber's famous thesis of the significance of "The Protestant Ethic" for modern capitalism. With this argument we are not here concerned, although it is related to the suggestion Weber made about science, which is of interest to us. With his knowledge of the great historical societies, Weber knew that men had always been more or less successful in their adjustment to the empirical world. He was convinced, however, that the religious values of the Graeco-Christian society were more favorable to the development of empirical science than those of the other societies he had studied. We have already mentioned some of these more favorable views and attitudes: the view that the natural and supernatural realms were separate; the view that God was rational and that the natural universe reflected his rationality; and the view that man could discover the rational order in the natural universe. These are not attitudes taken by all religions and all societies. Weber has himself shown, for example, that the view of the world held in Classical, Confucian China was different from that of the West; and the predominance of what Weber calls "the magic image of the world" in China helps to explain the lack of science in that society.[47]

What is of immediate relevance here is Weber's suggestion that Calvinist Protestantism, or what he called "The Protestant Ethic," was an especially favorable version of Christian attitudes toward the world for the development of science. Calvinism brought the great forces of Mediaeval rationalism into everyday life, thus stimulating empirical science, because Calvinist theology held it man's religious duty to order his "this-worldly" activity of all kinds, economic and otherwise, in the most rational fashion possible. Gradually, of course, over the centuries, this religious attitude has been secularized, until the purpose and justification of rational empirical activity is no longer immediately, but only indirectly, religious. In the seventeenth century, however, when the justification was still religious, the new Calvinist view of the world furnished a strong impetus to the growth of science. Rationality about the empirical world was now, paradoxically, enjoined by the supernatural sanction of Calvinist theology.

Weber's suggestion, made specifically in connection with the Calvinist Puritans of seventeenth century England, has been, as we have said, taken up by Robert K. Merton, American sociologist, and put to a careful empirical test, a test which we can only sketchily report here. First of all Merton made a detailed and quantitative study of scientific activity in seventeenth century England, using chiefly the papers in the *Philosophical Transactions* of the Royal Society as his evidence, and demonstrated that the number of Puritans who were active in science and the extent of their contributions to it are disproportionately greater than those of other religious groups and particularly of the Catholics. Merton also, incidentally, has collected statistical evidence to indicate that this disproportionate participation in science by Protestants as against Catholics holds right up to modern times for the European continent, as well as for Great Britain.[48] Of this differential, so far as it concerns the United States, we shall say more later.

Then Merton turned to the set of religious beliefs and attitudes which constitute Calvinist Puritanism, beliefs expressed in theological writings, in sermons, and in books of moral guidance for the layman. It is this set of beliefs, this thing labelled "The Protestant Ethic" for the sake of brevity, which makes the difference in the propensity to scientific activity. What are some of these beliefs? The Puritans held the view that man could understand God through understanding Nature, because God revealed himself in the workings of Nature. Therefore science was not antagonistic to religion but rather a firm basis for faith. They felt that since "good works" were a sign, if not a proof, of election to salvation, and that since one could glorify God through social utilitarianism, then science was good because it was an efficient instrument of good works and social improvements. And they valued reason highly because God had chosen man alone to possess it and because it restrained laziness and idolatry. The Puritans did not esteem the empirical world for its own sake but rather as the stage on which rational, orderly activity—so useful for science—was approved by God. The congeniality of these religious views for scientific activity is obvious. "The combination of *rationalism and empiricism* which is so pronounced in the Puritan ethic," says Merton, in summary, "forms the essence of the spirit of modern science."[49]

We have used Merton's study to show the influence of religious factors on the rise of modern science, but we may repeat that it is of general methodological interest in the study of the social aspects of science. It has the essential scientific virtue of showing a *direct and specific* connection between scientific activities in a particular place and time, on the one hand, and a carefully defined and isolated social factor, on the other. For example, Merton does not claim that Protestantism as a whole has this beneficial influence on science, but only Calvinism, and Calvinism only necessarily at a given stage of its development. And lastly, Merton, like Weber, does not claim that the relation between science and "The Protestant Ethic" was consciously intended by the Puritans of seventeenth century England. Social influences on science, like the reciprocal influences of science on society, we have already noted, are often most powerful just when they are unintended.

And here we may stop in our all too brief account of the historical evolution of science. There is, we trust, no further need to exemplify the six themes which run through this account and which could be as easily demonstrated in the history of science during the few centuries which succeed the period of which we have just been speaking, the period of the rise of modern science. We turn now, in the next chapter, to an account of the social and cultural factors which are most favorable to the maintenance of a high level of science in the world of the twentieth century. From now on we shall be more interested in the analysis of the nature and conditions of a fully-developed science than in the history of its evolution.

III

Science in Modern Society: Its Place in Liberal and in Authoritarian Society

AFTER OUR account of the historical development of science, we can now understand in just what sense it is true, as we science, we can now understand in just what sense it is true, as we so often hear, that science is unique in modern society. Our science is not unique in kind but rather in its extremely wide scope and in its high degree of development. Only in modern society do we find that peculiar combination of elements which has evolved out of earlier forms of empirical rationality and which is indispensable for science as we know it: — very highly generalized and systematized conceptual schemes; experimental apparatus which greatly extends man's powers of observation and control of data; a relatively large number of professional scientific workers; and, widespread approval of science in the masses of the population as well as in the elites.

This combination of elements, this science we know and take for granted, is not, however, random, nor is it inevitable or immutable. Recent events in Nazi Germany and in Soviet Russia have suggested that some parts, at least, of science may decline as well as grow, perhaps even be stifled altogether. Science, in short, is not only dependent on its environing society, as we have seen in the preceding chapter, but is more congruent with some types of social conditions than with others. This relationship between science and modern society has recently been noted by Professor Talcott Parsons. "Science," he says, "is intimately integrated with the whole social structure and cultural tradition. They mutually support one another—only in certain types of society can science flourish, and conversely without a continuous and healthy development and application of science such a society cannot function properly."[1]

Throughout the rest of this book we shall be making as detailed an analysis as we can of the relations between science and society. In this chapter, however, we want to take a broader view of these relations, a view which is the necessary propaedeutic to the finer analysis which follows. We want to isolate some of those relatively macroscopic social conditions which characterize the modern Western world in comparison with other kinds of society and which make possible a high level of scientific activity and advance. We shall speak, therefore, of such things as our cultural value of rationality, our highly specialized division of labor, and of the significance of these things for science. How such features of our society have evolved out of a large number of social changes covering many centuries, changes including those owing to science itself, will not concern us now. We are interested only in their present congruence with science, the congruence Professor Parsons speaks of.

The purpose of singling out these broad characteristics is not to describe precisely just what modern society is. What we want for the present is a kind of model, an "ideal type" as some sociologists would call it, of what this type of society is *in comparison with* other societies. This model is nowhere fully realized in the different societies of the modern world, but it is realized in greater and less degree. We can use our model, our set of characteristics of modern society, as a rough but useful measure of the *degree* of relative favorableness which these different modern societies present for the development of science. Specifically, examination will show that certain "liberal" societies—the United States and Great Britain, for example—are more favorable in certain respects to science than are certain "authoritarian" societies—Nazi Germany and Soviet Russia. We say that the latter countries are "less favorable"; we do not say that science is "impossible" for them. This is not a matter of black-and-white absolutes but only of degrees of favorableness among different related societies. Now ideological thinking about these matters does deal in absolutes; for example, it speaks of "the death of science in Nazi Germany." Science is harder to kill than that in the modern world. Such thinking will not carry us far toward a sociology of science which is useful for realizing our values as well as for genuine scientific understanding.

Every human society has a set of cultural values, a set of moral preferences for certain kinds of social activities as against their alternatives. Let us turn first to the system of cultural values which characterizes the modern world as against other societies, the values which realize themselves not only in science but in a great many other social activities. This is the set of deeply-rooted moral preferences which has made possible the uniquely high development of the science we know. This is the set of values we must maintain relatively strong if we wish to maintain science relatively strong. These values are not, of course, officially nor even informally codified, so the particular list we give here can only be offered as the consensus of numerous scholars and moral leaders who have tried to discover them. Any similar listing would, however, probably have a very large overlap with this one, especially when merely verbal differences were eliminated by close analysis. In any case, these are the values that are significant for science and other essential activities in modern Western society, even if it is difficult to draw them up in precise and final hierarchies.

One of the key cultural values we have to speak of is the value of *rationality,* and the congruence of this moral preference with science is obvious. We are not now referring just to the practice of rationality, for as we have seen that occurs in all types of society. By the value of "rationality" we mean the moral, the emotional, the "institutionalized" as the sociologist says, approval of that practice throughout wide areas of the society. This approval results in the critical approach to *all* the phenomena of human existence in the attempt to reduce them to ever more consistent, orderly, and generalized forms of understanding. Rationality of this kind is specifically different from what has been a predominant characteristic of all previous types of society, namely, the cultural value of "traditionalism." This value approves the acceptance of whatever exists, on its own terms, simply because it has always existed, without wishing to criticize it in terms of rational consistency and generality. The "rational bent" of modern man, which Thorstein Veblen was one of the first to compare with the habits of men in other societies, leads modern man to question the world in every direction, to analyze all that has been passed down to him merely by the "rule of custom." The modern world thinks the rule of reason more important than the rule of custom and ritual.

This value of rationality underlies much more than science in our society, although it is most strikingly manifested there, of course. For example, our economic activities can only be maintained in their present form because of the widespread diffusion of this value in the population. The moral norm for behavior in the economic area, that is, is a rationality of which efficiency in industry and a skillful orderliness in all affairs are the outward signs. When we praise "the spirit of free inquiry" we are referring to another aspect of our value of rationality. That spirit is chiefly exercised by the professional groups and especially by the scientists among these, but it is a cultural ideal in all social groups. Every man, we say, has the right to ask questions and to satisfy "himself," by which we refer to his reason. Indeed, this is not merely a right, but a duty. That is to say, there is a notably active quality about the value of rationality in our society. It requires man to strive for rational understanding and control of all his affairs by a perennially active effort, not just when events baffle or thwart him. In science itself, this spirit of rationality becomes an institutionalized self-generation of endless inquiry, ever novel and ever more general hypotheses. No realm of the world or of society is now immune to penetration by the active rationality prescribed by our cultural approval of it.

Inevitably, of course, as we well know, this active rationality comes into conflict with certain established habits and activities in society, for example, with the "sacred" beliefs of religion or with ancient economic mores. These other activities resist the "attacks" of rationality, sometimes violently, more often in our recent history by giving way slowly in adaptation to the corrosive effects of unbounded inquiry. We shall look into the sources of this resistance to rationality and into its significance for science more closely when we discuss, later on, the social consequences of science. What we need to remark at this point is that, on the whole, the relative strength of the value of rationality continues to prevail, despite counter-attack and resistance from some of the things it questions and criticizes. Especially as it is embodied in the structure and consequences of science, thus, active rationality is the source of the great dynamism which sets its mark on the modern world.

We need a term for another important cultural value of the modern world and we shall use the term *"utilitarianism"* for this purpose, even though it has certain connotations which we do not here imply and which we shall therefore specifically exclude. By the value of utilitarianism we mean that predominant interest modern man has in the affairs of this world, this natural world, rather than in other-worldly affairs such as supernatural salvation. This value is also obviously favorable to a high development of science. Modern rationality, in contrast to say the rationality of the Mediaeval world, is primarily rationality applied to the empirical phenomena of everyday life. In our discussion of the rise of modern science, we indicated that this everyday empirical rationality was derived in some part from the active interest in this-wordly affairs prescribed by The Protestant Ethic of Calvinism, an interest which has been so well analyzed by Max Weber. By now, however, this interest in mundane activities has become almost wholly autonomous, almost wholly based on secularized derivatives from the earlier religious interests, as well as on the consequences of other developments. But perhaps this partial source of utilitarianism in what were specifically religious interests should make it clear that the value of utilitarianism is *not* necessarily and invidiously "materialistic." There is no identity between materialism and utilitarianism, as some who have opposed this latter value maintain. Although materialism is a possible consequence of utilitarianism, so is an "idealistic" concern for the affairs of this world also possible. The evidences of idealistic utilitarianism are widespread in social reform and social voluntarism. The most vivid evidence of all for the existence of idealistic utilitarianism, however, may be found in science itself. This too we shall refer to again, when we outline the specific cultural values of science as an independent social activity.

The approval our culture places upon *universalism* constitutes still another value which has a special congruence with the maintenance of a high level of scientific activity. This value, derived from and still expressed most fundamentally in the Christian ideal of the brotherhood of man in God, has a secularized meaning in modern industrial society. It means that, in this society, ideally, all men are free to find that calling in life to which their merits entitle them. It means that each man's station in life is consequent upon his achievement in his calling, that is, in an industrial society, in his "job." Every man may compete for any occupational function and for any specialized position within the hierarchy of an occupation. Quite specifically, for example, any man who has the talent for the job of being a scientist and who has the desire to take up that job, has a social right to do so that is as great as that granted to all other men. As the American expression of the value of universalism has it, a man may become a scientist regardless of race, color, or creed. Moreover, once a man has become a scientist, he has the right to be treated by all fellow-scientists and by all fellow-citizens in terms of the universalistic norms which apply to all who have that job and that status. Where the value of universalism is realized fully, Jews and Negroes are not barred from science or any other occupation. And in science itself there is no "Catholic" science and no "Jewish" science and no "German" science. Universal science flourishes in those parts of the modern world where the value of universalism is most nearly realized.

Another cultural value which has great scope in the modern world in contrast with other societies is the value that we shall call *individualism*. By this value we mean the moral preference for the dictates of individual conscience rather than for those of organized authority. We have the libertarian conviction, derived in part, as is *ultilitarianism,* from Protestant theology, that it is our duty to seek the inspiration for all behavior in our own consciences. Modern man grudges the sway of organized authority in a fashion which is new among societies. This is an attitude which is most congruent with science, for science rejects the imposition of any truth by organized and especially by non-scientific authority. The canons of validity for scientific knowledge are also individualistic: they are vested not in any formal organization but in the individual consciences and judgments of scientists who are, for this function, only informally organized. Some of the resentment which scientists feel against so-called "planning" in science, as we shall see more fully later when we discuss this subject, derives from their individualistic fear that formally organized authority will be substituted for the informal judgments of peers in the control of science.

One last cultural value of the modern world seems to be important, and that is the value placed upon *"progress" and meliorism.* There is, in present-day society, a widespread conviction that the active rationality we have discussed earlier can and should improve man's lot in this world. This is coupled with a belief in and an approval of "progress" in this world, a progress which is not necessarily of a unilinear evolutionary kind, but which is somehow cumulative in the way in which science and rational knowledge are cumulative. This value, too, has its source to an important degree in Christian perfectionism and Protestant activism. And of course our moral preferences for "progress" and meliorism have a positive congruence with the essential dynamism of science. On the whole, despite localized resistances and hostilities in particular cases, modern society has been cordially receptive to the innumerable innovations fostered directly and indirectly by the advance of science. If it be hard to live with the instability and change that science makes a permanent feature of our society, and we all know how hard it sometimes is, still our approval of science as an agency of "progress" and meliorism makes us more willing to sustain this condition, to take the bad with the good.

These, then, are the values we find in the modern world that make it so peculiarly congenial to science.[2] We have already said they are differently realized in different societies, and we shall say something of how this is so in a little while. We have also to notice, right away, that this system of values is not rigidly constituted even in the societies which realize it most fully. There is always some ambivalence toward all cultural values in all societies. There is surely some ambivalence, for instance, in modern society toward such things as the value of rationality and toward science itself. Since it is in such ambivalences that important possibilities for social change arise, it is necessary to mark this lack of rigidity even in cultural values that are central in a society. Later, we shall consider in detail some of the ambivalences toward science which are connected with its consequences for modern society. Any sociological model for measuring societies, however roughly, must be a dynamic model.

In addition to the cultural values that we have picked out, there are in the modern world, in contrast to other societies, certain social conditions which are especially favorable to a high level of scientific activity. We refer to such things as a highly developed division of labor, a social class system which permits of considerable social climbing, and a political system in which the autonomy of many diverse authorities is respected. Here too, in our discussion of these social conditions, we shall be constructing a model which is nowhere fully realized but only in differing degrees in different modern societies. These social conditions, or social structures as the sociologist might call them, are particularly congruent not only with science but also with all the cultural values that are characteristic of the modern world. They are not, however, merely derivative from those values. The two kinds of things are somewhat independent of one another, for all their possible congruences. For example, social action in terms of cultural values may have consequences that destroy social structures. This is what the Nazis were doing,—weakening their industrial system by their espousal of the cultural value of emotional irrationality. Contrariwise, of course, changes in the social structures of a society have consequences for its cultural value. For instance, the increasing value we put upon "security" as against "freedom" in American society is in part a consequence of changes that have occurred in our economic system. Because of these reciprocal influences between the different parts of a society, we have to consider, as we here shall, both the social structures and the cultural values. We turn now to these characteristic modern social structures, starting with the occupational system.

In every society, whether it is small and non-literate or large and "civilized," there is some division of labor and some specialization of occupational function. The degree of this division and specialization, however, varies enormously among known societies. In its very simplest form, for example, the division of labor may differentiate only between the work functions of men and women, adults and children. But even in such a relatively undifferentiated structure, the special skill or knack which particular members of the group have in certain tasks is usually recognized, at least to the extent that such people are informally recognized as the leaders in group occupational tasks. Thus, among the Trobianders whom Malinowski describes, there is a "canoe expert" who has greater experience and skill in building outrigger canoes and who therefore is the leader in any enterprise of this kind. Now only very small groups have the simplest division of labor we have just pictured. In groups of even a moderately large size and of even a moderately large accumulation of wealth and technique, the division of labor segregates numerous specialized occupational roles. Such occupational roles are usually fused with other social roles in a way which is different from modern industrial society. They are, that is, typically not segregated from the family roles which the craftsman has. The craftsman's work or the farmer's work, for instance, is passed down from father to son and thence to the next generation. This close connection between family and job are symbolized in the work-place, for workshop and family residence in this other type of society are not separate, as they are with us, but in the same place.

Modern industrial society is quite different. It has carried the division of labor to an extreme degree of specialization which has been hitherto unknown in human society. For example, taking the United States as a case, the Dictionary of Occupational Titles prepared by the United States Employment Service of the Department of Labor consumes more than a thousand pages in listing the titles and descriptions of the different jobs which exist in this country. The Dictionary defines some 17,000 different jobs, and this is admittedly not a complete list. In the American textile industry alone, there are about 1850 types of specialized skills.[3] *Ideally,* moreover, these jobs are allocated on a basis which ignores differences of family connection. They are supposed to be distributive points in an occupational achievement system which is based on merit alone. This kind of specialized and family isolated occupational system is a late emergent in the history of human society and is fundamentally important for the successful functioning of an industrial type of society. It was for this reason that the Nazi attempt to re-establish "race" and family criteria for the assignment of occupational functions was a threat to its industrial system, however much the Nazis consciously may not have wished this consequence of their actions.

In the light of these variations in the division of labor that we have been describing, the occupational role of the scientist, let alone all its extremely specialized sub-divisions, is by no means a "natural" occurrence. Except within the last few hundred years, science has been very largely the by-product of occupational roles devoted to quite other tasks than that of the development of generalized conceptual schemes tested by technical observational devices. In doing his job the craftsman often produced substantial rational empirical knowledge, sometimes all unwittingly, sometimes self-consciously. But only in the modern industrial system, with its elaborate division of labor, is there a socially recognized and highly approved place for the "worker" whose job it is, and whose only job it is, to know science and to advance it. Indeed, such occupational positions do not appear full-blown until even later than the rise of modern science. We have seen that the great scientists of the sixteenth, seventeenth, and eighteenth centuries were typically "amateurs," or men for whom science was often an avocation, however passionate their interest in it. The men who then produced science often lived by other means, and they did their science as best they could, when they could. Benjamin Franklin was such a scientist, and a man of great scientific accomplishments.[4] If the "amateurs" were particularly fortunate, they might find a patron who admired science and who would therefore give them funds for research. Society as a whole laid out no clearly marked and generally approved careers for scientists. Not until the late nineteenth century, as we shall see later in some detail, is there a firmly established social basis for large numbers of scientists in the universities, industries, and governments of Western society. And in the twentieth century, so much is the occupational role of the scientist taken for granted and approved, that we may wonder at the need for pointing out that this was not always so. With its very many different types of jobs, its extreme specialization, and its internal organization into professional societies—of which, too, we shall say more later—, the elaborate occupational structure of science is now an essential part of the complex division of labor which is required by modern industrial society. This is as much true, we shall see, for a Communist industrial society like Russia as for a liberal industrial society like the United States or Great Britain. The continual advance of science now depends on this provision for a large number of occupationally specialized roles for scientific workers. Anything which diminishes this number and this specialization thereby potentially diminishes science.

The advance of science is in still another way related to the complex division of labor in modern industrial society. Not only are science and its products highly specialized, but an elaborate specialization of industry and technology is now required to use the production of science. In his very perceptive book, *Mechanization Takes Command,* Siegfried Giedion has recently shown that industrial technology of the kind we are familiar with is a function as much of certain kinds of social organization as of certain kinds of scientific knowledge.[5] The industrial assembly line, for instance, is an important social invention in the division of labor, and modern machine technology is impossible without it no matter how much scientific knowledge we have. Therefore, since science and technology are now extremely interdependent and fructifying for each other, both are fundamentally dependent upon the maintenance of that great division of labor which is so essential a characteristic of modern industrial society.

The type of class system which is more characteristic of the modern world than of other societies, the "open class" system as the sociologists call it, that is, a system in which a relatively large amount of social climbing is approved, is also especially congruent with the maintenance of science at a high level. This is because of the functions which social mobility has in society. That is, whatever the causes may be—and they seem to be in part genetic, in part social and psychological—the social elite of a society in any given generation does not entirely reproduce its successor in the next generation. This is true no matter what skills are required of the elite, whether they be military, administrative, scientific, or other kinds of skill. In every society, therefore, some form of social reproduction of the elite is necessary, and this is achieved through varying amounts and types of social mobility in different societies. If the channels of mobility in a society, for example, are nearly closed, the elite may fail to reproduce itself in sufficient numbers, with consequent harm to the effective functioning of the society.

The necessity for social reproduction of the elite group in any given generation seems to be as great for science as it is for any other activity, perhaps greater. Because of the highly developed and highly specialized abilities which scientists must have, the advance of science requires that it be a "career open to talent," one in which ability occurring in the lower classes may climb into the professional scientific classes. And so it has been very largely in the modern world. Science would soon stagnate if its functionaries were mostly mediocrities whose occupational positions had been ascribed to them on the basis of their family affiliations alone. Other particularistic criteria besides family would be equally perilous to science. No "race" or nationality group or class has a monopoly on scientific ability. For this reason, the Nazis hazarded a great deal when they excluded so-called "non-Aryans" from the profession of science. An open-class system, providing opportunity for all talent to express itself, is most congruent with the advance of science. Of course, the relationship between these two is reciprocal. For by providing in each generation a number of highly esteemed positions which are open to achievement, science performs an important validating function for an open-class society. We may say that where men must and can rise, notions of social and racial caste will have a harder time of it.

Science has another important connection with the open-class system in the modern world. Although many different motivations attract men to specific occupational careers, the degree of prestige in the open-class system awarded to any given career is an important differential element in the choices men make among the occupational alternatives that are open to them. In modern society, science has a high class prestige. The job of scientist ranks near the top, as we shall see later, in public evaluations of the scale of occupational possibilities. Indeed, social respect for science and its practitioners is widespread even among those groups where there is considerable ignorance of its nature and functions. We shall see that this same consensus is characteristic also of Russian society, despite its interference with the activities of particular scientists. In Nazi society, on the contrary, there was at least an ambivalence toward the prestige of scientists and even an hostility which greatly depressed their social position. Men are less attracted to science when its social prestige is lowered. On the whole, the high social status which scientists have in the modern world symbolizes public recognition of the social importance of their functions. No modern industrial society can afford to lower that status very much or neglect those functions.

Like a highly developed division of labor and an open-class system, the type of political system which does not largely centralize its authority also is particularly congruent with science. This "liberal" type of political system is, of course, a peculiar product of modern society in contrast to other types of social system, even though its incidence is partial even in the modern world. As we shall see more fully later, in the highly developed state to which empirical science has now arrived, its effective functioning requires a large degree of freedom from certain restrictive kinds of external control. Science cannot advance without a large amount of self-control, by which we mean control by the professional scientists themselves in their informal and formal organizations. This essential autonomy has, by and large, been granted to science in the modern world. Before the rise of modern science, this autonomy was incompatible with the hierarchical religious organization of the Church. More recently, threats to the freedom of science have come most often from hierarchical political organizations, notably in Nazi Germany and in Soviet Russia. The advance of science is hampered where scientific work is not judged by the canons of scientific activity but rather wholly by the political and social necessities of the authoritarian state. All modern societies do not now provide equally favorable political conditions for science.[6]

We have seen, in the discussion of our six themes on the social aspects of science, that the autonomy of science, like that of other social activities, is a relative and not an absolute one. Science never has been and never can be absolutely free of some control by other elements in the society, including, of course, the political element. The freedom of science is a matter of degree, a matter of specific forms of self-control. It will be our purpose to analyze these specific forms throughout this book, showing their functions for the advance of science. Such an analysis will, correlatively, indicate which kinds of control are not harmful to science. We cannot set science up against all the rest of society; the task of the sociology of science is to define their most fruitful type of interconnection.

Moreover, the relationship between science and the political system may change its specific forms, although not its general requirement of relative freedom for science. Such changes require adjustment based on understanding. In American society, for instance, although a high degree of autonomy for science continues, the relations between science and the political structure have been changing, especially during the last twenty years. As science has advanced and has grown in its social effects and social usefulness, the results of its advance have become political problems. The American Government has been continuously and on a fairly large scale concerned with the political problems of science since the Depression of the 1930's. Previously such a degree of concern had been limited to periods of war, although, as we shall see, the American Government has been somewhat involved with science for quite a long time. And since World War II, more than ever before, because of the relation of science to national defense and national prosperity, science and the Government have had increasingly close connections. This change has been remarked by President Conant of Harvard, who has himself been an extremely active and powerful participant in these relations. "Members of Congress and civilian officials of the Federal Government," he says, "have become involved in intricate questions which in large part turn on judgments about scientific and engineering problems. There can be no doubt that politics and science, once quite separate activities, have become intermeshed, and at times the grinding of the gears produces strange and disturbing noises."[7] Some of these new problems we shall consider in our discussions, first, of the place of science in the American Government and, second, the planning of science. Despite these problems, however, the necessary kind of autonomy for science seems to have been preserved in this country, most fortunately not only for science but for the whole society.

This, then, is the "ideal type," the model of the system of cultural values and social structures which would provide the most favorable conditions for science and its progress. In the degree in which one finds *all* these things—the cultural values of rationality, utilitarianism, universalism, individualism, and melioristic progress; and the social structures of a highly specialized division of labor, an open-class system, and a non-authoritarian political system—in that degree science flourishes in a modern society.

Before proceeding to a brief application of this model to certain "liberal" and "authoritarian" societies in the modern world, we must remind ourselves of the limitations of this "ideal type." The cultural values and social conditions that compose our constellation are of crucial importance for science, but they do not exhaust the factors in modern society which are in any way relevant to the success of science. At any given time, certain other conditions are also important. For instance, the matter of cultural values and social structures apart, it makes a great difference to science how much scientific knowledge has already been amassed in a given society. Thus, when Nazi Germany diverged from the value of universalism, say, it still made a difference that the earlier German society had built up a great corpus of scientific work which remained available to Nazi society. Similarly, of two societies with cultural values and social structures judged roughly the same against our model, the one with the more accumulated economic and natural resources, with the more literate, educated, and skillful population, and with the greater expedient awareness of the virtues of science as a form of power, this one will be the more favorable to the development of science. We leave these and other social factors out of account here, although a full analysis of any given society would have to include them. We shall refer to some of them in later chapters: for example, we shall speak of the importance of the inherited body of scientific knowledge in the process of scientific discovery and invention. Just now, however, we want to isolate those cultural and social factors which are often neglected, even though they are of strategic significance.

Perhaps it hardly needs to be said again, especially after all our explicit and incidental references to the situation in the United States, Nazi Germany, and Soviet Russia, that the "ideal type" of favorable social conditions for science which we have just constructed is more nearly descriptive of what we call "liberal" societies in the modern world than of those we call "authoritarian." Not *exactly* descriptive, of course, but still more characteristic of the nature of the "liberal" than of the "authoritarian" type of society. In the modern world, however, both these kinds of society have had at least some of the features essential for science: for instance, an industrial economy based on a highly specialized and highly rationalized division of labor. This should warn us against making absolute categorizations of the state of science on the basis of so rough a dichotomy as "liberal" and "authoritarian." Indeed, the two so-called "authoritarian" countries of which we shall speak shortly —Nazi Germany and Soviet Russia—, while not so favorable to science as the United States or Great Britain, are relatively unfavorable each for quite different as well as for some similar reasons. Indeed, in broad comparative perspective, taking whole society against whole society, Nazi Germany was probably a much greater divergence from our "ideal type" than is Soviet Russia today. These important differences between the two countries can be brought out by an analysis in the light of our discussion up to this point.

Just now we shall say very little about science in "liberal" society beyond the most general remark we have already made that it is most nearly favorable, among various types of society, to modern science. Throughout the rest of this book, we shall be examining in great detail science as it exists in one "liberal" society, the United States. We shall point out many respects on which it diverges from our model. But also, we shall see the relatively great congruence of that society with our "ideal type." For instance, we shall find that the values of its scientists and of its people are roughly similar to the cultural values of that "ideal type." We shall examine how its scientists are recruited through the mobility channels of an open-class society, especially through its colleges and universities. We shall look at the immense occupational specialization of science and of industry in the United States, a double specialization which is highly useful for scientific progress. And we shall describe the relatively autonomous, informally organized structure of the scientific profession and consider the functions and problems of that autonomy as against external political controls. We could do all of these things equally easily for another "liberal" society like Great Britain, but we shall not. Simply for the sake of convenience, we shall speak chiefly of the United States and only incidentally of Great Britain. Despite his own dissatisfaction with the state of science in Great Britain, Professor Bernal's book, *The Social Functions of Science,* is a good demonstration of the relation between the "liberal" character of British society and its excellent science. Leaving "liberal" society aside for the moment, then, we shall take up two different "authoritarian" societies.

Nazi Germany turned away, not completely, but in a degree

which had harmful consequences for its science, from three of the conditions which are component elements of our "ideal type": the cultural values of universalism and of rationality, and the political condition of relative autonomy for science. It is impossible to measure these harmful consequences at all precisely, even now that we have been able to discover something of what was happening in Nazi Germany during the '30's and during the war years.[8] While Nazi science was not utterly destroyed, as it was predicted it would be by some scientists in "liberal" societies in the first shock of their moral and emotional reaction to the Nazi revolution, still the damage done was quite considerable.

Consider, for example, the consequences of the Nazi denial of the cultural value of universalism. Positively, this denial meant the glorification of the particularistic virtues of "Aryan Germans" as scientists. As a result of this attitude, candidates for scientific teaching positions were required to meet certain "Aryan" standards of physical, moral, and "racial" fitness, standards which have no demonstrable connection at all with scientific talent.[9] And negatively, of course, the denial meant a violent attack on German Jewish scientists and on something that devout Nazis referred to as "the Jewish evil" (*jüdischer Ungeist*) in science. The general consequences of the Nazi violation of the universalistic right of German Jews to continue as scientists or to train for the profession are evident in the serious losses of scientific personnel that Germany suffered during the '30's. Here are some rough figures. Between 1933 and 1938, 1880 scientific men of first-class distinction were exiled from the universities of Germany and Austria. Professor Needham, the English biologist, has estimated that more than 25% of Germany's Nobel Prize winners were among the 18% of all the men of scientific reputation who were banished. By 1937, the number of students in the natural sciences at the German universities was only about one-third of what it had been in 1932.[10] Some of the special consequences of harm to German science should also be mentioned. By denouncing modern atomic and relativity physics as "the Jewish science," *par excellence,* the Nazis brought the whole subject so much into disrepute that it became unpopular at the universities. This, of course, was a great blow to the recruitment of new workers in the field and to continuing research.

The limitations placed upon universalism in Nazi society did not mean, however, that all social mobility into the profession of science ceased. But the attack on Jews and other "non-Aryans" destroyed a source of scientific ability that had hitherto been extremely important in German science. The roster of distinction in pre-Nazi German science had a great many Jewish names.

The German turning away from the cultural value of rationality was also, of course, and necessarily, only partial. It is easier for fanatic Nazis to recommend that Germans "think with their blood" than it is to run a complex industrial society on that precept. The esteem of irrationality probably was found more often in propaganda speeches than in everyday administration of the society. It was, to be sure, often found in the propaganda. Herr Bernhard Rust, Reichsminister of Education, for instance, said at the celebration of the 550th anniversary of Heidelberg University in 1936: "National Socialism is justly described as unfriendly to Science if its valuer assumes that independence of presuppositions and freedom from bias are the essential characteristics of scientific inquiry. But this we emphatically deny."[11] Probably the most harmful consequences to science from the proud irrationality of the Nazi leaders were unintended by them, rather than deliberately sought. Hitler, it has been reported, often countermanded the advice of his assistants, advice based on rational investigation and planning, in favor of "hunches." Sometimes these half-irrational insights led to success; sometimes to failure. For instance, because of his irrational desire for miracle weapons, Hitler was susceptible to wild and quackish notions about scientific possibility. In the modern world, hunches are a weak foundation for decisions of state. National policy has to be right more often than hunches allow; it requires the best available rational empirical knowledge. Hence the importance of a strong value on rationality in the leaders as well as in the followers of a modern industrial society.

The most directly evil effects upon German science came from the new political authoritarianism of the Nazi government. The great German universities, which had been the pride of the earlier society, were very quickly subjected to political control by the Nazis, who seem to have had an especial distrust for academic scientists. Not only were many "non-Aryan" professors dismissed, but those who remained were put under the authority of men chosen for their party loyalty rather than for their scientific accomplishment. As a result, charlatans sometimes competed with competent scientists for funds and apparatus for research. Political authority could override knowledge established by scientific research and validated by scientific peers. From 1939 on, for example, all scientific theses for the Ph. D. degree had to be submitted to official Nazi censorship. Even earlier, from 1935 on, attendance of scientists at any scientific congress either in Germany or in some other country was subject to the approval of the Science Congress Center, an agency of the Reichsministry of Propaganda, whose chief we have quoted above. When delegations were sent to congresses outside Germany, they went under an appointed leader, "chosen for his reliability as a member of the Nazi Party."[12] This is not, as we shall see later, a satisfactory degree of autonomy for science; this is not the way in which science can operate effectively in the modern world.

Despite all the harmful consequences from the three sources we have treated, German science was far from extinguished, if we judge by its performance before and during the recent war.[13] The particularistic attack on the Jews and political authoritarianism harmed some parts of science much more than others. For instance, though the Nazis had disparaged higher learning and pure science, "they may have strengthened the position of technicians and persons engaged in development."[14] Research in the German Air Force was much better than that in the Army because Goering, head of the Luftwaffe, "before and during the war, employed pre-Nazi officials of known ability in technical capacities, even to the extent of having General Milch (a man of Jewish blood) as wartime head of the Air Forces Technical Office."[15] Apparently there was always some conflict among and within Nazi officials between a pragmatic attitude toward the power of science and a moral disapproval of it for its rationality. Even the anti-rationalist Nazis, therefore, toward the middle of the war, under pressure of the impending loss of the war, favored a heavy subsidy for scientific research. It was, however, too late. Nevertheless, the summary picture we receive from accounts of German science during the war is that it was still very good science on the whole, although much less good than it had been twenty years before. It was a science living on the fat of basic research accumulated in the pre-Nazi period. It is hard to predict what would have happened in the future if Germany had won the war.

Now this variability in the quality of German science and this perpetuation of certain kinds of science at a fairly high level in Nazi Germany raise important sociological questions to which we can only suggest answers here. If German science was not utterly destroyed by the Nazis, but only seriously weakened, how long does it take to "kill" science? Indeed, can it really be extinguished in a modern industrial society? Prabably not, and probably it cannot even be weakened beyond a certain point in such a society, as the Nazis seem to have discovered in the middle of the war. Their change of heart, based on expediency rather than on moral preference, to be sure, indicates that the sheer necessity for science in a modern industrial society may eventually cause a reaction against the social conditions which are harmful to it. In the short run, and what the short run is we cannot say at all precisely, a great deal of harm can be done to science by those who despise it and put it under too much political control. In the long run, another imprecise notion but also an important one, science could even be destroyed entirely. But this could only occur on payment of a very heavy social price, the loss of the ability to maintain an effectively functioning industrial society. This is not to say that even the harm done in the short run is tolerable to a modern industrial society. Especially in a world where powerful national industrial societies compete in peace as well as in war, the short run may be the significant time-span for social calculation. We cannot prove that this was so for the Nazis, but it does seem very likely that they greatly weakened themselves by changing those social conditions which are essential to a progressive science.

Soviet Russia is another modern "authoritarian" industrial society where we may find the harmful consequences of excessive political control of science. Here, in all accuracy, we must say that we still know very little about the details of these consequences; certainly we know even less than we do about what happened in Nazi Germany. For one thing, until quite recently, which is to say until after the late war, political control of science in Russia seems not to have been excessive, despite the great pretensions to such control that have existed "on paper."[16] The freedom of science in Russia was once much greater than it is now. For another, not enough time has elapsed since the imposition of direct political control—as in the Communist party approval of Lysenkoism over neo-Mendelian genetics—to weigh carefully the harm that will be done to Russian science. But if the experience of Nazi Germany counts for anything, and if relative scientific autonomy is as necessary as we know it to be, then the Russians will have to pay a price for their direct political control of at least some, and perhaps all, parts of their science. For direct political control of science in Russia—for example, political control of the *particular theories* that should be held in a given scientific field—seems to be spreading from biology to other fields—most recently, to physics.

The spread of political authoritarianism in Russian science is all the more striking when we note how congruent with science in general other aspects of Soviet society are. In contrast with Nazi Germany, the Russians had not abandoned universalism, although of course there have been great violations against this value in practice, particularly against "resistant bourgeois" and "enemies of the state." More recently, unfortunately, an increase in strong Russian nationalism has diminished somewhat their support for social universalism, but this is an attitude in which they are not alone in the modern world. What is bad, though, is that they are now more and more speaking of "Russian science" and "bourgeois science," as if science were not an international unity. The Russians have not, however, lessened the emphasis the modern world places on the cultural values of rationality and utilitarianism. Indeed, in these two respects, the Russians are in the main stream of development of Western society, and if anything, they have pushed their approval of rationality and utilitarianism to an extreme.[17] "More than one observer," says President Conant of Harvard, "in the course of the last two decades has been impressed by the deep concern for science manifested by the Kremlin."[18] The Russians have glorified science, quite self-consciously, as an instrument of social revolution and social planning, and they have given it great support both directly and through the enlargement of their whole educational system. During the first five-year plan, for example, beginning in 1929, the Russian government expanded the number of scientific academies, research institutes, research workers, and "aspirants," or scientific recruits. To take only the planned expansion of the latter, the aspirants: in 1930, there were to be 1,000; in 1931, 2,600; in 1934, 6,000; and in 1935, 4,000. This shows the order of planned increase in all scientific activities.[19] "According to an authoritative statement," says Leontieff, "in 1942 there existed 1,806 research institutes: 452 devoted to fundamental research in natural sciences and mathematics; 570 in various fields of industrial research; and 399 in agricultural research."[20] Their whole social creed, the Russians constantly remind us, is not just "Marxism," but "*scientific* Marxism." This creed, they feel, has the rational solution for every empirical physical and social problem, and therefore it is considered to be the most potent instrument for realizing the value they attach to social progress and social meliorism. Hence the veritable cult of science among all classes in Russia today. Hence also the great Russian striving for rational mastery over Nature in man's interest. The idea of this struggle (*borba,* as the Russian word has it) permeates all Soviet activities.

The changes that have occurred in several parts of the Soviet social structure have also been of the kind that is favorable to the development and maintenance of a high level of scientific activity. The aspect of the great industrialization of Soviet society that most interests us here is the vast increase in the specialization of scientific and other occupational roles. The specialization in science and the specialization in industrial technology have been mutually fruitful, as they always are in modern industrial society. This transformation of Soviet society has been possible, of course, only because of the practically unlimited social mobility which has occurred, only because of the selection of scientific and other talent from all groups in the society, wherever it may be found. Although the amount of this social mobility now seems to be decreasing somewhat, as inevitably it had to, it still seems to be the equal of what occurs in American society, and it is highly conducive to the recruitment of competent scientific personnel.[21]

Yet against all these changes that are favorable to science, the change toward greater political authoritarianism over science works its contrary effects. To quote further from President Conant's shrewd observation: "That a wholehearted acceptance of science by politicians can lead to the curtailment of the work of scientists seems to have been clearly demonstrated" in Russia.[22] How these opposing influences are to be weighed against one another, no one can yet say. But we may predict certain possible consequences. Since science is to some extent an interconnected web of activities and theories, the several parts stimulating or retarding one another, political control of even a few areas of science may diffuse its harmful effects to other areas. The deterioration of Soviet genetics has already spread its influence to biochemistry and neuropsychiatry.[23] More immediately, perhaps, political intervention in any given scientific area undermines the stability of established scientific control in that area. Scientific fanatics and quacks—men like Lysenko in genetics—take over when political authority demands what competent scientists cannot conscientiously give it—particular substantive theories or results "on order." Where scientific authority is endangered or destroyed, competent men fear to take a position on scientific theory itself, for the demands of political authority are changeful and make almost any position insubstantial. Even further, in such a situation, competent men avoid a scientific career altogether. In all societies, men seek a *relatively* "safe" occupational career. A "flight from science," perhaps only to the more applied branches of scientific technology, as was the case in Nazi Germany, may be one of the unintended consequences of the extension of political control of science in Russia.

Not one, but two different pressures in Soviet society are apparently responsible for the recent extension of political control over science. The first pressure is the one which is the more commonly remarked, the need of an authoritarian political system to include within its direct control *every* activity in the society in order to have effective power over *any* activity. Analysis here runs as follows: in order to keep the Soviet educational system "in line," for example, the Communist Party must impose its organizational control even into the far reaches of "pure" science. The second kind of pressure that is evident in Soviet society is perhaps a more satisfactorily specific and identifiable one. That pressure comes from the great need the Russians have for immediate "results" from *all* activities in the society, science included as much as industry. The pressure for usable scientific theories in their agriculture and in their industry tends to force them toward demanding such theories from science, or, in the case of apparently competing theories, to choose that theory which is more immediately useful. This seems to be one of the causes for the Communist Party's support for Lysenkoism in genetics. Lysenko promises immediately useful scientific theories for the improvement of agriculture; for example, that he can create stable genetic changes in plants and animals and thereby produce breeds and species "on order" as required by Soviet society.[24] The Russians are probably not completely unaware of the necessity to let "pure" science have its head to some extent, but their determinist philosophy and the immediate needs of their agricultural and industrial system, including planned advances of course, push them towards the sacrifice of "pure" for "applied" science.[25] Sometimes the push delivers them into the hands of scientific incompetents. In this perspective, Lysenko is not unique but only a prototype of the men who may come increasingly to wield authority over Soviet science.

The case of Russia, then, like the case of Nazi Germany, demonstrates how necessary for modern science is the whole constellation of cultural values and social structural conditions that we have included in our "ideal type." A violation of any of the values or an alteration in any of these social conditions will have harmful, if not necessarily fatal consequences for science. We may repeat the quotation from Professor Parsons with which we introduced this chapter: "Science is intimately integrated with the whole social structure and cultural tradition. They mutually support one another —only in certain types of society can science flourish, and conversely without a continuous and healthy development and application of science such a society cannot function properly."

IV

The Social Organization of Science: Some General Considerations

WE HAVE just been examining the relative congruence of certain macroscopic features of liberal and authoritarian societies with highly developed science. Now we move from the more to the less macroscopic, and even to the relatively microscopic, in order to inquire further into the actual functioning of such science in liberal society, the type of society which, we have seen, is somewhat more favorable to the continued advance of science. Henceforth we shall be concerned primarily with science in American society, and that means we shall need to look more closely at the social organization both of science itself and of American society as well. In this chapter we shall take up some general considerations about the social organization of science, and in the next several chapters we shall try to see just how these manifest themselves in the actual scientific activity of the United States. Most of what we have to say is, of course, generally applicable to other liberal societies in the modern world; allowances can easily be made for minor differences which do not alter the nature of the essential relationships between science and liberal society.

First of all it has to be understood that science is, like all socially organized activities, a moral enterprise. Science, that is, can not be construed simply as a set of technical rational operations but has to be seen also as a set of activities devoted to definite moral values and subject to clear ethical standards. The proximate a-morality of the individual scientist is only possible, we shall see, because of a more ultimate, relatively absolute morality of science as a whole. Sometimes the moral ends of science are obscured by concentration on the a-moral means to those ends, but moral values are always present in the everyday working practices of scientists, however unconscious of them some scientists may be. Indeed, it seems to be characteristic of the values of science, as it is of other deep-felt moralities, that they should mostly remain implicit. Only seldom are these values made explicit; even less often are they codified by some official organization of science. Ceremonial gatherings and times of crisis are the chief occasions when the values of science are paraded. We must look chiefly to such occasions and to the reflective self-examination that occurs sporadically among mature and wise scientists if we wish to see the morality that governs science.[1]

We wish to discover this morality because it sets so many conditions for the social organization of science with which we are now concerned. If we look to the sources we have already mentioned, and to writings about "the scientific attitude," especially when they come from the pens of experienced scientists, we can find a very large area of agreement about the integrating moral components of science. There is even a certain tendency among those who hold the scientific values to glorify them as the special moral virtues of scientific activity alone and to ignore their connections with the larger values of liberal society, those values which we described in our last chapter. Such moral provincialism is often characteristic of the enthusiasm of the morally devoted; it is not peculiar to science. It is essential to the sociological understanding of science, however, to see how very largely the morality of science coincides with the more general morality of liberal society, and how, even where it differs somewhat, the difference is only made possible by that same more general morality. This mutual harmony of the more general and more special moralities is, in fact, positively functional for both: for if there are moral appeals that science can make to other participants in liberal society, appeals that find a responsive audience, so also the achievements of science furnish moral exemplifications that strengthen the fabric of values that binds the whole society.

In a certain sense, perhaps, science is morally typical of liberal society. Nevertheless, though coincidence and congruence between the two moralities are what we shall now examine, there is no simple identity. The special problems of the social organization of science in liberal society require certain separate as well as common moral norms.

Let us look first at some common values and their relationships in science and society. Faith in the moral virtue of rationality, we have seen, is one constituent of the "ideal type" of liberal society. Such faith is, if anything, strongest in those sections of liberal society where science prevails, for approval of the utmost powers of reason is a central moral value in the social organization of science. Here the interaction is quite clear: the general liberal faith in rationality probably receives its strongest reinforcement from the continuing achievements of science. When men waver about the virtues of reason, science is a powerful and insistent reminder of its worth. We can see how this occurs even among the scientists themselves. "As a participant in the most successful intellectual enterprise of the human race to date," says Professor Percy Bridgman, the scientist "is in a peculiar position to have won the conviction that not only is there no substitute for using one's mind, but that the problems which confront us are soluble, and soluble by us." And, he goes on, "if physicists will only make others see their own wider vision, their ultimate influence will far transcend that of any possible technological contribution."[2]

The faith of the scientist in rationality is peculiarly intense, and necessarily so, for it must be strong enough to persist in the face of great difficulty and repeated failure in scientific work. This intense moral conviction, no matter what the difficulty, has been beautifully expressed in a sentence by Einstein which has been engraved on the fireplace of a room in Fine Hall at Princeton University: "Rafiniert ist Herr Gott, aber boshaft ist Er nicht." We may translate the sentence freely: "God—Who creates and is Nature—is very subtle and difficult to understand, but He is not arbitrary or malicious." All things are possible to Reason in the form of high developed science, Einstein believes. And we need only observe that this belief in the virtue of rationality is not simply an act of intelligence; it is also a testament to moral conviction.

But this belief in rationality, this devotion to the "truth" which the rational conceptual schemes of science can discover, however sacred it may be to the moral community of scientists, is not, of course, a belief either in Absolute Truth in the large or in any particular truths. Scientific morality holds only that it is worth endless human striving to attain those inherently provisional and approximate statements of truth that make up the substance of science at any given historical moment in its course of development.[3] It is this conviction that science is necessarily ever-changing, ever-growing, that resists as immoral any attempt to fix Truth once and for all by tradition or by political authority. This is a "hard" faith, but it maintains itself in the social organization of science. In its theoretical substance, then, but not at all in its moral values, science is "eternally provisional." The great French physiologist, Claude Bernard, has spoken of this in his classic work, *An Introduction to the Study of Experimental Medicine.* "My theories," he says, "like other men's, will live the allotted life of necessarily very partial and temporary theories at the opening of a new series of investigations; they will be later replaced by others, embodying a more advanced state of the question, and so on." Scientific theories, he says, "are like a stairway; by climbing, science widens its horizon more and more, because theories embody and necessarily include proportionately more facts as they advance. Progress is achieved by exchanging our theories for new ones."[4]

Because of its intense faith in rationality, science appears to be characteristically "critical," even to men imbued with the general values of a liberal society. This is so because the morality of science tends to drive it into *all* empirical areas. We now see, however, that the inspiration for this activity is not the fear of empirical problems which so often underlies "critical" investigations. The goal of science is not attack, but understanding, and this goal is based on the moral value that all things must be understood in as abstract and general a fashion as possible. If the goal is sometimes misconceived as attack, and it is, especially when the problems are social phenomena, this is not the intention of the conscientious and self-controlled scientist. Nevertheless, this misconception causes troubles for science, and we shall have to see, later on, what these troubles are.

There is a value in science which is ancillary to faith in rationality and which is more important in science than it is generally in liberal society, although it is far from uncommon also in applied sciences like medicine and even elsewhere in the occupational sphere. This is the value scientists set upon emotional neutrality as an instrumental condition for the achievement of rationality. Science approves of emotional neutrality not primarily for its own sake, and certainly not for all social activities, but insofar as it enlarges the scope for the exercise of rationality, and its power as well. Emotional involvement is recognized to be a good thing even in science —up to a point: it is a necessary component of the moral dedication to the scientific values and methods. But in the application of those techniques of rationality, emotion is so often a subtle deceiver that strong moral disapproval is placed upon its use.

This is not to say that strong emotions are entirely absent in the relations among scientists themselves. If they are less frequently expressed than in other social activities, if they are more controlled, still, enthusiasms and fervent conviction, vehement attack and violent defense sometimes occur in science as they do elsewhere. Pasteur, to take but a single, notable example, engaged in a number of what his biographer calls "passionate" controversies on problems of theoretical interest:—with Liebig on the germ theory of fermentation; with Pouchet and Bastian on spontaneous generation; with Claude Bernard and Berthelot on the intimate mechanisms of alcoholic fermentation; with Colin on anthrax of chickens; with Koch on the efficacy of anthrax vaccination; and with Peter on the treatment of rabies.[5] "There were also," says Dubos, himself a scientist, "conflicts involving priority rights, or those arising simply from the clash of incompatible personalities. Whatever the cause of the argument, scientific or personal, Pasteur handled with the same passion those whom he believed to misrepresent the truth, or to be prejudiced against him." Pasteur was "jealous of his right to his discoveries," and, "he wanted to be one of those for whom cities are remembered." Pasteur is not at all unique in the annals of science. In all their specialized fields, scientists have been something more than bloodless automatons. The ideal of emotional neutrality, however, is a powerful brake upon emotion anywhere in the instrumental activities of science and most particularly in the evaluation of the validity of scientfic investigation. Pasteur was a great scientist *despite* his emotionality, and he has always been a greater hero to the public at large than to the scientific public. The scientific ideal runs more to the dispassionate genius like Claude Bernard, Pasteur's contemporary who is so much less well known.

Another value of science which is connected directly with the larger morality of liberal society is the value we have called "universalism." In science all men have morally equal claims to the discovery and possession of rational knowledge, as in liberal society all men have equal claims to life, liberty, and the pursuit of happiness, and as in the sight of the Christian God all men have equal rights to charity and grace. The universalism of all three of these spheres, we have seen earlier, is not unconnected in origin or present basis. Scientific truth is not conditional upon the social or personal qualities of the individual scientist. Regardless of his race, creed, or color, every contributor to the corpus of scientific theory becomes a member of the "community of scientists and scholars," sharing in its privileges and esteem in proportion to the merit of his achievement. This is a community of moral partners, and its reach is beyond the national group; science is international, it is universal in its ideal. The notion of an "Aryan" or "Russian" science is therefore abhorrent to science. In the "brotherhood of science," as some of its members speak of it, tolerance deriving from universalism is an absolute moral virtue. No scientist can safely harbor any preconception that certain new ideas in science are necessarily good or bad. It is required of scientists that they tolerate the possibility at all times that any new idea, no matter what its social source, may be a useful idea in science, that is, that it may subserve the essential scientific task of constructing better conceptual schemes. Hence the necessity to tolerate the whole universe of people, because scientific contributions have already been made by men of all sorts, and all sorts of men have the potential ability to be trained to make such contributions.

One last value which is essential for the social organization of science and which is shared with the larger liberal society is the value we have called "individualism" and that expresses itself in science particularly in anti-authoritarianism. The moral approval of "freedom" is not, of course, a charter for caprice in scientific activity; indeed, science is one of the most disciplined of social activities. But the discipline of science is that which the individual imposes on himself because of his faith in rationality and his moral convictions about the proper methods of realizing that faith. This discipline is supported mainly by the similar moral convictions of the individual's scientific peers and also by the numerous informal types of social control which express these moralities. The individual scientist obeys the moral authority of his peers because they share his values. All other authorities in science he rejects as immoral. Freedom of investigation, in both direction and extent, limited by no authority alien to the absolute morality of science, is the ideal that scientists cleave to.

A striking case of the internal anti-authoritarianism of science has been described by the mathematician, Leopold Infeld.[6] Infeld was invited by Einstein to collaborate and he did so for three years at the Institute for Advanced Study in Princeton. During this time Einstein was interested in building a bridge between the gravitational and quantum theories, and Infeld soon felt skeptical that this could be done. "It seems presumptuous," he says ,"that I would dare to differ with Einstein on any subject, but I know that there is nothing so dangerous in science as blind acceptance of authorities and dogmas. My own mind must remain for me the highest authority." Accordingly, he told Einstein of his doubts and objections. "Looking back on it now," he continues, "I must admire the patience with which Einstein treated my objections. When we started he was far ahead of me in this problem and I had difficulty following him. But he was never impatient; he came back many times to the same explanation of ways and methods, considered all my doubts seriously until I had absorbed the principal idea." The pattern of anti-authoritarianism was as much respected by Einstein as by Infeld; both recognized the moral duty of the scientist to follow the authority of his own judgment.

We come now to certain ideals of the social organization of science which are somewhat different from the dominant patterns of liberal society as they exist today, although these ideals are important in some other areas than science proper and could even some day become the dominant moral values for the whole society.

The first of these is what we may call the value of "communality."[7] Where liberal society as a whole values private property rights in scarce goods, in science such rights are reduced to the absolute minimum of credit for priority of discovery. Beyond this minimum, all contributions to the fund of scientific knowledge and conceptual schemes are community property, accessible to all competent members for use in the community's interests. In science, if anywhere, the utopian communist slogan becomes social reality: "From each according to his abilities, to each according to his needs." In the community of scientists, all scientific peers have the right to share in the existing knowledge, because many men have contributed to it in the past and all are potential contributors to it in the future. It is in the sharp light of this value of "communality" that secrecy in science becomes an immoral act. The man who takes from science according to his needs has the moral obligation to publish any new discoveries he builds upon the goods that the community has lent him. Secrecy in science is also, of course, dysfunctional in other than moral terms.[8] Secrecy shuts scientists off from the work that their fellows have already done and deprives them by that much of the necessary materials of their own work. It also eliminates what we shall see is essential to all scientific innovation, namely, the informal discussion among scientists of new work and new ideas. Much innovation in science is the product of the slow increment of smaller novelties. Such increments occur everywhere in science as fruitful consequences of informal discussion of work-in-progress by busy scientists. A scientist shut off from personal contact with his colleagues by the requirement of secrecy in his work, even though he have access to their publications, is always handicapped in some small degree and sometimes his handicap is insuperable. We shall speak of this again in our analysis of the social process of invention and discovery.

Only in times of extreme crisis, when defeat in war threatens not only science but liberal society itself with destruction, will scientists accept the restriction of secrecy. And even then they accept it only in limited areas and only temporarily, as "a dire necessity" in the interests of sheer survival in order eventually to restore the customary morality of science. In times of peace, the requirement of secrecy—say by the military, as has happened recently in the

United States—arouses moral conflict in many of the scientists who participate in research under this condition. Academic scientists seem to disapprove of such peacetime secrecy more than do their peers who are in the employ of the Government and of industry, but only 47% even of the last group, the industrial scientists, approve without qualification of secret research.[9] The anti-secrecy norms of science are powerful and pervasive.

Closely connected with the value of "communality" in science is the ideal that has been called "disinterestedness" or "other-orientation" by Talcott Parsons.[10] This too is a moral ideal that is not universally expected in our society but is essentially limited to science and to certain areas within the liberal professions, most notably in scholarship and medicine perhaps. In the larger society, men are expected to be "self-interested" in their occupational activities, "self-interested" in the sense that they serve their own immediate interests first, although any such activity may of course indirectly conduce to "the greatest good of the greatest number," and, indeed, it is part of the ideology of *laissez faire* society that it will *necessarily* do so. But in science a different moral pattern prevails. There men are expected by their peers to achieve the self-interest they have in work-satisfaction and in prestige through serving the community interest directly, and this is done through making contributions to the development of the conceptual schemes which are of the essence in science. The different moral ideals, it should be clear, are not matters of typical differences in the personalities of scientists and other men. In science, as in other social activities in liberal society, for example, in business, men seek the generalized goal of "success." In science, however, the rules of the game for achieving success are different: they enjoin the individual to serve himself only by serving others. Without "disinterestedness" as one of the rules of the game in science, it is unlikely that the value of "communality" with regard to scientific innovations could prevail. If too many men should draw upon the scientific theories held in common only to use them for their own immediate purposes, for example, in the service of their personal power rather than in the service of science itself, then the community property would cease growing and thereby lose its essential scientific characteristic.

These two moral ideals, of "communality" and "disinterestedness," are not, we may repeat, limited to science. In liberal society, as in other types, these are values which are always at least in some small degree relevant to men's everyday behavior. All of us are expected to make some direct contributions to the community welfare. In science, however, the scope for these values is much larger than it is in other kinds of social activities.

We may see this greater scope in the attitudes which scientists have toward the patenting of their discoveries. There is a difference here between so-called "pure" and "applied" scientists—a difference about which we shall say a great deal in a little while—but for the former at least, patents are an immoral infringement on the common property of science. "Applied" scientists in industry conform more nearly to the business ideals, of necessity. For "pure" scientists, patents are accepted, like secrecy in research, as necessary evils under certain special conditions, for example, when some scientific discovery should be protected in the immediate public interest and should not be published for possible exploitation by commercial enterprises. This is the case typically with biological and chemical discoveries which have medical applications that are immediately apparent to the research scientist. In such circumstance, the morals of science hold that a scientist may permit his discovery to be patented in the public interest, but only on condition that he himself receive no direct financial benefit from such a patent. The University of Wisconsin Alumni Research Foundation, for example, was specifically organized to develop in the public interest the patents on the use of ultra-violet rays to enrich foods with Vitamin D as an anti-rachitic agent, these patents being based on the discovery of a Wisconsin professor, Harry S. Steenbock, who has not himself profited from the returns on this patent.[11]

These several values, then, shared in greater or less measure with the other social activities of liberal society, are what constitute science as a moral enterprise. The morality of science is not always immediately apparent, and perhaps least so when it is most effective. When the social organization of science is operating successfully, under the control of these values so deeply held and widely diffused among working scientists, the controls are taken for granted. Only when violations occur within science itself, or when non-scientific authorities seek to impose new values on science, do these moral codes become more apparent. In some recent times of crisis for the whole society, as in war and economic depression, scientists have become more self-conscious of their values, and their leaders have sometimes proposed that these values be generalized to the whole society. Perhaps nowhere is the morality of science more evident than in such proposals. Thus, in a recent presidential address to the American Association for the Advancement of Science, the members are urged to use the resources of science "as a means of enriching and strengthening the spirits of men and breaking down the barriers which now divide them." This address speaks of the "high adventure with the universe which science is." It recommends that "the brotherhood of science" be used to promote the universalistic brotherhood of man. Another fundamental aspect of the morality of science is expressed in the phrase, "this ministry of science to mankind."[12] These are deep-felt convictions; science, like all other social activities, has one of its foundations in a set of moral values.

But of course, we say immediately, science is more than just the set of values we have sketched. There are other determinants of science as a social activity. We have said that scientists act somewhat differently in different kinds of organizations, in the universities and in industry, say, with regard to such matters as secrecy in research and the patenting of discoveries. The nature and purposes of the kinds of group in which scientists work have a significance which cannot be neglected; we need to specify what this general significance is. One convenient way of doing this is now to consider intensively the differences between "pure" and "applied" science, differences to which we have already several times loosely referred. This approach is especially convenient because it will also give us an opportunity to look at some other important aspects of science.

We have already dealt at length with science as a set of conceptual schemes and with science as a set of moral values. We have also just now suggested the importance of the different social groups in which scientific work is organized. And we have also earlier referred to the matter of the personal motivations that individual scientists may have in their own work. To understand the distinction between "pure" and "applied" science we shall have to see how all four of these variable aspects of science—conceptual schemes, values, social organization, and personal motivations—are implied in the distinction and why each aspect has to be dealt with separately as well as in relation to the others.

Probably the essential reference point for any distinction between "pure" and "applied" science is the significance of generalized and systematic conceptual schemes for all science. These, as we saw in our first chapter, are the primary cumulative component of science. In this regard, "pure" science may be defined as science which is primarily and immediately devoted to the development of conceptual schemes, such development including their extension, revision, and testing, an inherently endless process of establishing provisional "truth." Those who have this aspect of "pure" science in mind often refer to it as "basic" or "fundamental" science, a recognition of the importance of conceptual schemes for scientific progress. "Applied" science, on the other hand, in this regard, is science which is devoted to making conceptual schemes instrumental to some other social purpose than that of the pursuit of conceptual schemes as ends-in-themselves. Much "applied" science in the past has rested on relatively empirical, low-level conceptual schemes, indeed, on notions and assumptions that could hardly be generalized at all. This is still true even in modern times; here is the realm of a great deal of "cut-and-dry" rule-of-thumb technology in highly rationalized industry. Much useful knowledge in the photography industry, for example, is of this relatively empirical kind.[13]

This aspect of the distinction between "pure" and applied" science is, of course, an analytical one. Both kinds of science, in this sense, may be involved, and very often are so involved in any given concrete program of scientific research. Both kinds of science occur in different types of scientific organization.

A second important dimension of difference between "pure" and "applied" science consists in the moral values which are expected in different kinds of scientific activities. All the ideals we have described—rationality, universalism, individualism, "communality," and "disinterestedness"—are expected of "pure" science, although as we have seen and shall see further, these are not without their limits even here. The limits on some of these ideals are characteristically greater in "applied" science. Perhaps even the limit on rationality is somewhat greater in "applied" science, since this is a virtue which must be maximized for the development of conceptual schemes, and "applied" science is often more "traditional" in its methods than is "pure" science. Certainly the limits are usually greater with regard to universalism, individualism, "communality," and "disinterestedness;" but some "applied" work in science may be as much devoted to the universal good of mankind as is "pure" science, it may be as anti-authoritarian, as "communal," and as "disinterested." We have already given an example of this kind of "applied" work, namely, medical research. Such a case indicates very clearly the necessity of keeping separate these two aspects of science—that of conceptual schemes and that of moral values. They cross-cut one another and must not be assumed always to go together in the same way. The development of conceptual schemes may be limited by nationalistic particularism; and "applied" science may be in the interest of the universal community.

"Pure" and "applied" science in both of the first two respects we have considered also vary independently with the type of social organization in which they are carried on. The development of conceptual schemes and the full realization of the values of "pure" science are typically found in the universities and colleges of liberal society. In contrast, in industry and in government research organizations, these purposes and values do not so much prevail. In government research, for example, "disinterestedness" extends only to the national community, at best, and not to the whole world, as it does in the university ideal of "pure" science. And of course the social organization of private industry requires each enterprise to maximize its own gain from all activities, scientific or otherwise, however much it may hold a "service" ideal and ultimately contribute to the general welfare. Such national and private limitations on the "communality" of science are not, to be sure, absolute, but they are greater than those that occur in university science.

There are, then, these typical differences in purpose and values set by these different kinds of social organization. The chief dwelling of "pure" science is in the university and of "applied" science in government and in industry. But in actual fact, some of both kinds of science—considered either as conceptual scheme or as moral value—can be found in all these places. Much science in the university is at least incidentally "applied," and more of it is explicitly so. Where medical schools and biology departments overlap in the university, for example, there will be found considerable "applied" research in the direct interest of human health rather than of conceptual schemes. Similarly, in industry and in government, much "pure" or "basic" research often has to be carried out in order to have theories to apply to the solution of problems which are set by the restricted interests of those organizations. Especially in some modern industries which rest on "basic" science—e.g., the chemical, radio, and electrical industries—there is provision in research departments for some theoretical scientific research, as well as for more immediately "applied" work. That is why two winners of Nobel Prizes for discoveries in "pure" science have come from American industry: C. J. Davisson of the Bell Telephone Laboratories and Irving Langmuir of the General Electric Company.

In every case, however, in the not too long run—as we shall see more fully in our chapter on science in industry—the "basic" research of industry has as its purpose some application in the immediate interests of the enterprise which subsidizes it. To think otherwise would be to ignore the social purposes of industrial organizations. The directors of industrial scientific research groups are well aware that their activities are subject to the same institutional imperatives as all other industrial activities, well aware that they must lead to the maximization of profit in the not too long run. Dr. C. M. A. Stine, Vice President in charge of research for the Dupont Company, with long and successful experience in his profession, has spoken of the "implied monetary motive for fundamental research in industry." Fundamental research in his laboratory, he says, "is not a labor of love. It is sound business policy. It is a policy that should assure the payment of future dividends."[14] It was Dr. Stine himself, subject to this industrial imperative, who was able to persuade Dr. Wallace Carothers to come from his research on high polymers at Harvard to work on the same problem for the Dupont Company. It was as a result of Dr. Stine's "sharp business foresight" in hiring Dr. Carothers for just this research that "nylon was deliberately forced into being."[15] Dr. Stine and his employers were willing and able to wait almost ten years for their investment in "pure" science—that is, as conceptual scheme—to pay off in application to the manufacture of the synthetic fiber, nylon.

In still another way types of social organization are relevant to the distinction between "pure" and "applied" science. It is sometimes believed that "pure" research can only be carried on by individual scientists working alone or in small teams, and "applied" research only in large-scale, hierarchically organized groups. As we shall see again later, these are perhaps typical differences between "pure" research in the university and "applied" research in the government and industry, but in point of fact both these modes of organization of research are found in each of these areas. The universities have organized large research hierarchies to exploit such research aids as the atomic cyclotrons and electronic mathematical calculators. Private industrial research organizations are glad to have a few of their men work by themselves on relatively "basic" as well as "applied" problems.

We have said that, in the long run, "basic" research in industry is also "applied" research. So also, of course, although in the somewhat longer run, is research done in the university, for all research ultimately has some application, whatever the more immediate purposes for its development. This is true even of that allegedly most "pure" of all scientific auxiliaries, higher mathematics. A recent account of the wide range of usefulness of mathematics in industry cites H. M. Evjen, mathematician in the geophysical research department of the Shell Oil Company, on this point. Mr. Evjen says, "Higher mathematics means simply those branches of the science which have not as yet found a wide field of application."[16] The author of the account himself adds, "The routine operation of our industrial system today, therefore, involves the use of transcendental equations, matrix algebras, Heaviside operational calculus, probability functions, analysis situs, and other mathematical systems and devices previously known only in advanced academic circles and dismissed by practical men as pure theory."

Yet the different time perspectives between "pure" and "applied" research, while not absolute and somewhat overlapping, must be carefully distinguished, because the differences between short run and long run purposes in science are essential to the social organization and the advancement of science. However much "pure" science may eventually be applied to some other social purpose than the construction of conceptual schemes for their own sake, its autonomy in whatever run of time is required for this latter purpose is the essential condition of any long run "applied" effects it may have.

Now perhaps we can see why it cannot be asserted, as it sometimes is, that "pure" science is that which has no "social consequences," as the phrase has it, and "applied" science that which has such consequences. If all science ultimately has consequences for some other social purposes than science itself, as we said in one of the "themes" in Chapter Two, then the distinction between "pure" and "applied" science on this ground can only be made on the basis of the relative length of time which elapses between scientific activity and its social consequences. The elapsed time is typically greater in the case of "pure" science; but it need not be, as the atom bomb bears witness.

The fourth and last of our variable aspects of the distinction between "pure" and "applied" science is that of the personal motivations of individual scientists. All that we have said up to now is probably sufficient to indicate that the distinction between the two is certainly not wholly a matter of the motivations of the individuals who carry on different kinds of scientific research. Yet some "common sense" and moralizing discussion errs in just this fashion.[17] "Pure" scientists are said to have "better" motives than "applied" scientists. Now there is no firm way of determining the correctness of such allegations, since there is very little evidence at all about the personal motivations of scientists. "Pure" science may have some characteristic attraction to men of one or more particular personality types; and "applied" science, on the other hand, may appeal to some still different personality types. It seems probable, however, from what we know in general about the lack of any fixed relationship between personality type and occupational role, that there is a very large range of overlapping in the motivations of men in different kinds of scientific work.

We can, at least, find testimony to this in what some scientists themselves say about the matter. The very same scientist, says

Professor George B. Kistiakowsky, professor of physical chemistry at Harvard University, can "enjoy" both kinds of research. "I feel," he says, "that there is no great difference between so-called applied and so-called fundamental from the point of view of the investigator. During the war years I enjoyed working on applied problems of explosives and some other things just as much as I had enjoyed the work without practical purpose which I had previously done at Harvard and hope to do again."[18] From the other side of the fence, an industrial physicist has recently defended his colleagues against allegations of "inferior" motives. "We in industrial physics," says John M. Pearson of the Susquehanna Pipe Line Company of Philadelphia, "find among us the whole range of human attitudes toward science just as is found in the university."[19]

In any case, as we can conclude from our preceding discussion of this matter, we do not need in the first instance to call up considerations of personal motivations. Typical differences in the kinds of social organization in which scientific research is carried on are at least as important in making distinctions between "pure" and "applied" science. Provision for the different kinds of science depends, therefore, upon the maintenance of the appropriate types of social organization. Together with the influences on personal motivation from the larger liberal society, these types of social organization—say, the university, industry, and government—provide adequate mechanisms for establishing and controlling the requisite kinds of motivation for scientific work. Just how this is done will be the main subject of our next several chapters.

Finally, we have to note that "pure" and "applied" science always have an important influence on one another, whether they are concretely separated, or not, in the same or in different types of social organization. Indeed, they are necessarily mutually dependent, for not merely does "pure" science provide new theories for social application, but these applications in turn furnish instruments and conditions for the easier advance of "pure" science. Their connections says President Conant, are "symbiotic" and "tight-knit."[20] The growth of science requires that "pure" and "applied" research never be too sharply isolated from one another. A great danger in misunderstanding the nature of the two, and their relations, is that it may lead to such a strict and harmful segregation.

V

The Social Organization of Science in American Society

THE SOCIAL ROLE of the scientist, taken in the light of that conception of the essential nature of science which we have described in Chapter One, may be said to have three different functions: to develop conceptual schemes, to train other people how to develop conceptual schemes, and to apply conceptual schemes to the realization of various social purposes. These are of course analytic distinctions, and in the concrete role of any given scientist we may well find them intermingled. In American society, we have said, these different functions are typically performed by scientists in three different types of social organization: —the university and college, industrial research groups, and Government research groups. The university performs primarily the first two functions, to develop new conceptual schemes and to train new scientists to develop them. Industry and government have usually performed chiefly the third function, to apply the conceptual schemes developed in the university, but they have also made some independent development of conceptual schemes, and their training function is not unimportant. Among these three types of social organization there are close, interdependent relations, arising not only out of the necessary interdependence of the three functions they perform but also from the concrete overlapping of these functions in each of them. All three groups make important contributions to the development of science in American society. As the President's Scientific Research Board has put it, "The programs of all three are important to the national welfare. Governmental policies must be shaped in recognition of the importance of this research triangle."[1] In our next three chapters we shall consider in turn and in detail each of these three types of social organization, their special structures and special problems. In the present chapter we want to take up some matters that apply to all three: the public prestige of the American scientist, his social rewards, and his work satisfactions; the professionalization and specialization of the scientist's role; the pattern of coordination and control in American scientific activity as a whole, and some of its problems; and, some basic facts about the present size of American science and about the social characteristics of American scientists. These are all general considerations which set the stage for the special performance of the university, industrial research, and government research alike.

The social role of the scientist, like all other roles in society, is subject to evaluations by the public-at-large as well as to self-evaluations. The two types of rating are, of course, intimately related, and also they are relatively harmonious in an integrated society; otherwise the roles would not be filled nor filled successfully. The social rewards of the scientist in public prestige, in money income, and in other honorific symbols are largely an expression of these public and self-evaluations. They constitute a body of social goods which help, together with immediate work satisfactions, to attract men to the role of scientist and to maintain the appropriate moral sentiments and individual incentives for a scientific career. These are general sociological propositions whose specific relevance for the social organization of science in America we may now examine.

First of all, public prestige. Our commonsense impressions that the prestige of scientists in the American occupational hierarchy is very high have recently been substantiated by a reliable empirical study. The sociologists, Professors C. C. North and Paul K. Hatt, have investigated public evaluations of different American occupations by means of a survey of a representative sample of the national population.[2] According to the findings of the North and Hatt study, when Americans are asked to rank 90 different jobs in their society, they give the highest rankings to positions which have two characteristics: highly specialized training and a considerable degree of responsibility for the public welfare. Americans seem to feel that the role of scientist has both these characteristics. Let us look at some of the high rating scores and see just where scientists are placed.

Public Ratings of American Occupational Roles

1.	U. S. Supreme Court Justice	96
2.	Physician	93
3.	State Governor	93
4.	Cabinet Member in the Federal Government	92
5.	Diplomat in the U. S. Foreign Service	92
6.	Mayor of a large city	90
7.	College professor	89
9.	Scientist	89
10.	U. S. Representative in Congress	89
11.	Banker	88
12.	Government scientist	88
13.	County Judge	87
15.	Minister	87
16.	Architect	86
17.	Chemist	86
19.	Lawyer	86
20.	Member of the Board of Directors of a Large Corporation	86
21.	Nuclear scientist	86
23.	Psychologist	85
24.	Civil Engineer	84
27.	Owner of a factory that employs about 100 people	82
28.	Sociologist	82
30.	Biologist	81

Inspection of these relative rankings shows that the role of college professor, which of course overlaps that of scientist, and the role of scientist itself stand pretty high. The generic role of scientist ranks somewhat higher than the specific specialties within science, perhaps because attitudes are less mixed toward the more abstract than toward the more specific conceptions of the role. The more abstract conception may evoke public liberal values unmixed with ambivalent feelings; the more concrete conceptions may, on the other hand, evoke some of the unfavorable attitudes toward the particular social consequences of certain scientific discoveries. For instance, nuclear scientists stand a little lower than scientists in general, perhaps because of their connection with the atomic bomb. Among the several different scientific specialties, as we might expect from their relative degree of maturity, the physical sciences rate slightly higher than the biological. The biologists themselves are somewhat aware of this lesser esteem, and they have expressed concern about the "lack of public appreciation of the contribution of the biological sciences" to the recent war.[3] The social sciences, at least so far as psychology and sociology represent them, also stand fairly high, indeed surprisingly so. In the last chapter of this book we shall look a little deeper into public evaluations of social science. One last point about these occupational ratings. Professors North and Hatt report that there is considerable consensus among their sample on all these ratings, but that Americans on the higher educational and economic levels rate the professional occupations slightly higher than do Americans on the lower educational and economic levels. This would seem to indicate that those who have directly experienced a liberal education or some scientific training are somewhat more imbued with the values which support science and also more aware of its functional importance in liberal, industrial society.

Now we may ask, How do these public evaluations express themselves in the "more tangible" form of money income. There is, of course, in American society as a whole, no necessary and fixed relationship between the public prestige of a job and its social reward in the form of money income. This is certainly true for the earnings of scientists. The evidence we have permits us to say that scientists do earn a higher average income than most other occupational groups, but that very few individual scientists command the extremely high salaries that occur in some areas of the American occupational system. Only a very few men, and these are the scientist-administrators in industrial research groups, earn even more than $20,000 a year.[4] A small proportion of all scientists earn over $10,000. Those who work in academic institutions are less well paid than those employed by private industry and government. The following figures, taken from answers received in a Fortune Magazine poll of scientists, indicate the relative earnings of scientists in universities and colleges, in government, and in industry:—

	Academic	*Government*	*Industry*
Under $2,000	8%	—	1%
$2,000—$ 4,000	20%	10%	10%
$4,000—$ 6,000	33%	35%	31%
$6,000—$ 8,000	18%	32%	24%
$8,000—$10,000	10%	16%	15%
Over $10,000	11%	7%	19%[5]

The individual is interested not only in the maximum salary he can achieve but also in the pattern of its increase during his career. In science, salary ceilings do not come early, as they typically do in relatively unskilled American occupations. The typical pattern, like that in most professional careers, is rather one in which advancing age, achievement, and experience bring slow, small, but steady increases in salary. In the Fortune poll already mentioned, for example, 65% of the men aged 25-35 were in the $4,000–$6,000 salary group. Of the men over 45, however, 25% earned more than $10,000. The income for the successful scientist, therefore, while not extremely high, reaches what most Americans would consider a fairly comfortable level in middle and older age.

The figures we have given lump all scientific specialties together. They can be compared with some others compiled from the earnings of a single group, the chemists, by the Committee on Economic Status of the American Chemical Society. In 1941, just before the war but when employment opportunities were good, the median annual income of the members of the chemical profession was $3,364., 50% earning less than this amount, 50% more. The lowest 10% had incomes of less than $2,000., the lowest 25% incomes of less than $2,500. Only 25% had incomes higher than $5,000., and only 10% higher than $8,000. Among the older members, those with more than 40 years' experience, 25% earned more than $9,694.; and 10% of this group earned more than the very high salary of $19,200. These are, of course, pre-war figures. During the war the same Committee found in a study made in cooperation with the United States Bureau of Labor Statistics that the income of chemical scientists had increased from 14 to nearly 80%.[6]

Although scientists are in general, as we shall show later, a fairly contented occupational group, they are not wholly satisfied with the social rewards they receive in the form of public prestige and money income. A survey of attitudes among scientists made for the President's Scientific Research Board found that a majority of scientists felt that "money and prestige rewards are less than they should be."[7] The feelings on this score were general and showed little direct relation to actual income level. This feeling and its generality would suggest not that scientists are an embittered occupational group but rather that they are much like all other American occupational and income groups, who feel that they could get along comfortably if only they had an income greater by ten per cent.

Perhaps all this talk about money income seems to imply that it has precisely the same functions in scientific circles that it has in other occupations. This is not so. In the business groups, for example, money income varies over a very much greater range than it does in science, and it is much more directly a symbol of one's relative occupational status. This can be seen even among those scientists who are attached to the business groups: the salaries of industrial scientists cover a greater range and go much higher than do those of their academic or government colleagues. Within each of the types of social organization in which science is carried on, the prestige of jobs and the salaries for them are roughly correlated; but there can be no precise comparison between, say, academic jobs and industrial jobs in terms of money income. High money rewards are considered a more suitable incentive to achievement in the business world. In university science, money rewards are not supposed to furnish a primary incentive to achievement, and there is some concern lest a system of monetary rewards displace the motivation for making discoveries from its proper goal to the goal of merely getting the money symbols of such achievement. The values of "communality" and "disinterestedness" in science, as we have seen, discourage the elaboration of finely graded invidious distinctions based on money income.[8]

These different honorific functions which money income serves in science and in industry are perhaps reflected in the different appropriate "style of life" for the two groups. Symbols of "success" for business achievement are the things which large amounts of money can buy, like expensive homes, fine cars, and costly clothes for wives. Among scientists, such symbols are considered inappropriate even when inherited income may make it possible for some few scientists to procure them. Competition among scientists is restricted to scientific achievement; "pecuniary emulation," as Veblen called it, is morally barred.

The most appropriate symbol of achievement in science is a man's job, the relative prestige of jobs depending somewhat on general public evaluations but much more on the evaluations that are made by professional colleagues. As we shall see later, there is a handful of American universities whose scientific professorships award the highest prestige in their fields to the holders of these positions. In industry, too, the research organizations of some companies have much more prestige than others and those who hold jobs in these groups share in the prestige. Neither in the academic or industrial groups, however, are the ratings of jobs firmly and finally fixed. Instead, they fluctuate somewhat as the achievements of the present incumbents vary in the esteem of the relevant professional group. In science, recourse to the test of achievement is quick.

Because it is so quick, and because he knows how few of his colleagues may be competent to judge his specialized work, the individual scientist values very highly both the informally expressed and the formally manifested opinion of his colleagues. In science, as is true everywhere, it is hard to get reliable evidence on the operation of that most powerful means of social control, informally expressed opinion, even though everyone knows of its existence and feels its effects. It is somewhat easier to get more formal expressions of esteem, such as elections to the offices of professional societies and awards of prizes for distinguished scientific accomplishment. In this respect, therefore, let us consider the matter of prizes and awards in science. We could take as our examples the very large number of these for which national and local groups of American scientists alone are eligible and about which one may read every week in the news columns of the magazine, *Science,* the journal of the American Association for the Advancement of Science which is read by practically all working scientists. Instead, to show the generality of the system of prizes in science as a whole we shall take the Nobel Prizes, which are awarded without regard to nationality. In the relevant respects these are typical of the nature of prizes as a symbol of prestige in science.

The committees which participate in the award of the Nobel Prizes in science are largely made up of men who themselves have already demonstrated outstanding scientific accomplishment, men whose opinion is therefore particularly highly valued. Not only may no direct application for a Nobel Prize be made, but also not everyone is allowed to suggest suitable candidates. The list of those who may nominate candidates includes members of the Swedish academy that actually awards the prizes—in physics, in chemistry, and in physiology and medicine—, certain professors in Scandinavian universities, former Nobel Prize winners, and selected persons of scientific distinction in other countries throughout the world. This latter group of selected persons is appointed for twelve months only, so there is a new selecting committee each year. This helps eliminate favoritism and also serves to incorporate the representatives of new developments in science. To assist in the selection of the 1949 prize in physics, for example, 237 scientists outside of Sweden, including 42 in the United States, were asked to propose candidates. It is, of course, an honor just to be appointed to the nominating committee, since only distinguished men are so chosen. The greatest honor of all, however, certainly one of the greatest that can come to any scientist, is to be awarded the prize itself. A Nobel laureate is honored wherever he goes in the world of science and usually elsewhere as well.

The Nobel Prizes illustrate also another typical characteristic of scientific distinctions, namely, that they are usually deferred rather than immediate. In the whole history of the science prizes, only once has an award been made for a discovery that was announced in the previous year. That was the prize in physiology and medicine awarded jointly to F. G. Banting and J. J. R. Macleod in 1923 for the discovery of insulin, which they and their co-workers, C. H. Best and J. H. Collip, had announced in 1922. Perhaps only in medicine could the significance of a scientific discovery be so immediately apparent. Ordinarily, the Nobel Prizes are awarded for achievements that are five to ten years old. The winners are mature scientists: the average age of the winners in physics has been 46 years, in chemistry, 49, and in physiology and medicine, 54.[9]

In addition to the social rewards of prestige and money income, immediate work satisfactions are an important type of incentive for satisfactory performances in scientific as in other occupational roles. Here we have some evidence from the study conducted in 1947 for the President's Scientific Research Board, a study which looked into the work satisfactions of a sample of 567 scientists in the universities, in industry, and in the United States Government.[10] This group made up the best possible sample that could be obtained under somewhat unsatisfactory conditions. As one might expect from the importance of the value of rationality in scientific work, intellectual satisfactions are those which the individual scientist rates most highly in his work and in his career. These intellectual satisfactions include "understanding the way things work, pioneering in the unknown, and exercising the creative impulse."[11] A strong secondary source of satisfaction is the social value of the work these scientists do. "A majority say the social contribution is a matter of concern to them, and that they feel their work contributes to the welfare of mankind."[12] This is an expression of that scientific value we have called "disinterestedness."

Work satisfactions are an important basis for choice among the various types of organization in which these scientists could carry on their occupation. "Opportunity to do the kind of work you want to do the way you want to do it is named first by each group of scientists (university, industry, government) as a basis for deciding which type of organization is most satisfactory."[13] On this score a very large proportion of these scientists are fulfilling their wishes. A majority say that "they are doing the work for which they are best fitted, have freedom of action to try out their ideas, and have opportunity to advance their professional competence." There is little variation in these attitudes, either within the three groups of scientists or among scientists of different income or age levels.[14] Despite this actual work satisfaction among scientists in all three types of organization, a majority of all of them taken together consider the university the most satisfactory type of organization in which to work. Industry ranks second and Government is a low third. We shall see the significance of these relative rankings later. The university is most highly valued because of its "freedom from restriction." Industry offers the satisfaction of "seeing tangible, practical results." And Government is attractive because of its "unlimited facilities and resources" for research.[15] Each work satisfaction has been at least sufficient to recruit adequate numbers of scientific workers for the different types of organization.

In sum, then, in the words of the Report, American scientists declare themselves to be comparatively well satisfied in their work with regard to the criteria which they themselves describe as paramount."[16] Much the same rough result is expressed in the answers of scientists to a Fortune Poll question, "If you had it to do over again, would you choose the same line of study?". The answers:

	Academic	*Government*	*Industry*
Yes	91%	86%	84%
No	9%	14%	16%[17]

Lest these problems of rewards and incentives seem peculiar to American scientists, let us take the comparative perspective for a moment and see how the same situations occur in the social organization of science in Russia. So far as public evaluations of Russian scientists go, we have no public opinion studies, of course, but if we may judge from other kinds of evidence, then the prestige of science and scientists is very high indeed in Russia. The British scientist Eric Ashby reports an extreme admiration of scientists in Russia, a public admiration so great and so widespread that it seems to him to be almost "hero-worship" for the living and "lay-canonisation" for dead scientists. He thinks, further that this attitude of "deep respect for science and scientists" helps to attract "the brightest minds in Russia." Great publicity is given in newspaper accounts of the many fine awards to scientists and, as a result, "the ambition of many a young Russian is to be a scientific research worker." There are even public festivals in honor of scientists and their discoveries, much as have we honored such men as Edison by issuing commemorative postage stamps.[18]

All this has a most familiar ring, and so too does the system of more tangible rewards which Russian scientists receive. The greatest honor and largest financial rewards accrue to the scientists who are elected to the several national and all-Russian academies of science. These men, the number of whom has been greatly enlarged in recent years, receive handsome salaries and, equally important, get such perquisites as better housing and extra rations of food and clothing, the right to buy scarce consumer goods like autos, certain income tax exemptions, and access to facilities for vacation and travel. There are also pensions for these men and for their widows and children. These pensions are reported in the newspapers, along with other special awards given to the families of men who achieve distinction in the occupational sphere. Such material rewards are not limited to men working in the fields of "pure" or "fundamental" science. So far as patents on inventions are concerned, for example, the situation in Russia now seems to be a great deal like that in Great Britain and the United States.[19] The Invention Act of 1941, replacing earlier legislation enacted first in 1931, defined the conditions under which patents and "author's certificates" may be granted and royalties paid thereon to inventors. The title, "author's certificate," incidentally, appears more suitable to the Russians than the more capitalistic one, "patent," although the two things are sociologically the same. As a result of the 1941 legislation, the career of the professional inventor was made legally what it had been in fact before, a select one in which the practitioners receive well above average income, food, clothing, and educational advantages for their children.

The most important reward of all, perhaps, both for the prestige they carry and the income they bestow, are the Stalin Prizes, which have a great similarity to the Nobel Prizes. These are also awarded in other fields than science, and the top prizes carry a very large sum of money. In 1943, for example, the physicist, Peter Kapitza, won a Stalin Prize paying about $30,000. for his discovery of superfluidity in helium. This is a discovery of great practical importance because it makes possible a much cheaper method of producing liquid oxygen and, consequently, great savings in the process of reducing ores in the metallurgical industries. Each year about $1,200,000. is awarded in Stalin Prizes for the "pure" natural and social sciences alone. There are also sixty prizes for inventions in "applied" science, including ten first prizes of about $16,000. each, twenty second prizes of $8,000. each, and thirty third prizes of $4,000. each.

So far as immediate work satisfactions are concerned, we have no direct information on what they are for the Russian scientists who work in the universities, in industry, and in government research laboratories. We may safely assume, though, from the general similarity of the various social incentives that are offered to Russian scientists to those that prevail in more "liberal" societies that the different kinds of work satisfactions they have are also much the same. The same, that is, so long as there are no intrusions of political authority into the social organization of science to the extent of imposing some particular scientific theory. We saw, in Chapter Three, that such intrusions have become more frequent in Russia recently. Because of the importance for scientists of such work satisfactions as "freedom from restriction," regardless of the society in which they find themselves, it is very likely that Russian scientists are now a much less contented as well as less effective group than they were formerly, before the recent increase of direct political control.

We may return now to our discussion of some general aspects of the social organization of American science. We have been speaking as if professional and specialized scientific occupational roles had always existed, but we have seen, in Chapter Two, that this is not so. We saw there that in the seventeenth and eighteenth centuries, when relatively mature science first emerged in Western European society, scientists were not only few in absolute numbers but also chiefly amateurs, that is, men whose main occupational role was something other than that of being a scientist. The amateurs of these early times were, to be sure, the equals of later-day professionals in their enthusiasm for science and very often in their actual competence. For example, Benjamin Franklin, who is almost the prototype of the distinguished amateur in science, made contributions to the theory of electricity which earn him an important place in the history of physics. When we speak of "amateurs" now, we are speaking of a kind of social role, not of personal devotion to science and not of the level of technical competence. The course of development in the social organization of American science, like that in roughly similar "liberal" societies, has been an evolution from small numbers of amateurs to large numbers of professional, specialized workers. We shall now trace this evolution, and some of its consequences, as it has occurred in American society.

We have seen that the early amateurs in science joined together in societies to provide themselves with a common meeting place for work and discussion. The first of these—the Royal Society in England, the *Academie des Sciences* in France, the *Accademia del Cimento* in Italy—were founded in Europe in the sixteenth and seventeenth centuries.[20] The first American society of this kind which is still in existence was the American Philosophical Society, founded in 1743 by Benjamin Franklin. The several sciences were all much more lumped together in those days, under the title of "natural philosophy," and amateurs usually tried to cover the whole field. Not until the nineteenth century did both professionalization and specialization occupy any sizeable part of the province of science, and these developments were much more retarded in the United States than they were in Europe. The "general indifference to basic research displayed in the United States during the greater part of the 19th century" was reflected in the slower professionalization of science in this country than in contemporary Europe.[21] We often forget how recent a development large-scale professional science is. For example, not many of us know that the term "scientist" itself was unknown until the nineteenth century, when it was deliberately coined by the Reverend William Whewell, Professor of Moral Philosophy in the University of Cambridge.[22]

The pattern of increasing professionalization is clear. In the first half of the nineteenth century it begins to appear in American colleges and in the Government itself, which at this time hires its first few full-time scientific employees. With the accumulation of the first large American fortunes in industry and commerce, more financial support is available for professional scientists, because the new capitalists were willing to endow positions for full-time scientific teachers in the colleges and even to endow whole sets of such positions in scientific schools. For example, the Lawrence Scientific School at Harvard, established at this time, was endowed by Abbott Lawrence, the pioneer New England manufacturer of cotton textiles; and the Sheffield Scientific School at Yale was similarly benefited by Joseph E. Sheffield, a Connecticut canal and railroad magnate. In the second half of the century, professional science was continuously enlarged, and now large-scale industry also provided a few jobs for full-time scientific workers.[23] In the twentieth century, as we shall see in our next three chapters in some detail, there has been a vast increase in the number of professional scientific positions in the universities, in industry, and in the Government.

Unfortunately, we have few statistics to illustrate even roughly this pattern of increasing professionalization in American science. The following table, which goes back to the year 1876, shows the increasing number of Ph.D.'s in *all* fields of scholarly activity, and it is, therefore, only a very rough measure of the increase in the number of men whose full-time occupational role is that of scientist.

Year	*No. of Ph.D.'s Conferred*	*No. of Conferring Institutions*
1876	44	25
1890	164	—
1900	342	—
1910	409	38
1920	532	44
1926	1,302	62
1928	1,447	69
1930	2,024	74
1935	2,649	84
1937	2,709	86[24]

The rapid acceleration in the number of professional scientists during the twentieth century suggested by this table is more reliably demonstrated in the following figures of the number of scientists listed in *American Men of Science,* the who's who of American science:

1903 — 4,000
1910 — 5,500
1921 — 9,500
1928 — 13,500
1938 — 22,000
1944 — 34,000[25]

We are now concerned only with showing the pattern of increase in the ranks of American professional scientists. It is a pattern which is roughly similar for that in other "liberal" societies. Later we shall give some figures which describe the absolute numbers of professional scientists which now exist in the United States.

This professionalization of science is, of course, only in part a result of the internal changes in science itself. It is in part also an aspect of the increasing professionalization and specialization of the whole occupational structure of American society. Just as science has made a greater division of labor possible, so, reciprocally, the increasing division of labor in American society has opened up large numbers of jobs in science. Science now has the increased stability which derives from its being an essential and regularized career in an integrated occupational structure.

Increased and still increasing specialization is another aspect of modern American science which comes equally out of changes internal to science itself and out of changes in the larger occupational system. Norbert Wiener, professor of mathematics at M.I.T. and the author of *Cybernetics,* has described the pattern of change here and its present circumstance quite vividly. "Since Leibniz," he says, "there has perhaps been no man who has had a full command of all the intellectual activity of his day. Since that time, science has been increasingly the task of specialists, in fields which show a tendency to grow progressively narrower." In the nineteenth century, he says, if there was no Leibniz, at least there was a Gauss, a Faraday, a Darwin, men whose knowledge and work compassed a whole large subdivision of science. "Today," however, "there are few scholars who can call themselves mathematicians or physicists or biologists without restriction. A man may be a topologist or an acoustician or a coleopterist. He will be filled with the jargon of his field and will know all its literature, but, more frequently than not, he will regard the next subject as something belonging to his colleague three doors down the corridor."[26]

Nowhere is the specialization of modern science more evident than in the hundreds, the thousands of different technical scientific journals that are published in the United States. The world over, including a great deal of overlap and even identity, of course, the number of professional scientific journals amounts to no less than forty thousand.[27] As a result of this proliferation of journals for each scientific specialty, specialists sometimes complain that they can manage only to keep up with the work in their own narrow field. There are some devices for alleviating the task of "keeping up" with other specialties and with general scientific interests, such as publications of abstracts of papers and general science journals. But one of the important problems of contemporary science, in the United States and elsewhere, is to maintain fruitful relations among the multitude of specialized scientific disciplines. Scientific conceptual schemes are both generalized and abstract, we have seen; specialization makes possible greater abstractness, but sometimes it hinders the greater generalization of scientific theory to which it should contribute.

Both increasing professionalization and increasing specialization are reflected in the changing types and growing numbers of American scientific societies. We may get some sense of the amateur, generalized interest of the early societies from the following description by Cotton Mather of a "philosophical" or scientific society that met in Boston in the seventeenth century: "A Philosophical Society of Agreeable Gentlemen, who met once a Fortnight for a Conference About Improvements in Philosophy and Additions to the stores of Natural History."[28] Today there are thousands of local and national professional scientific societies in the United States, and these represent hundreds of scientific specialties.[29]

These professional organizations in American science are of three kinds, the different types representing characteristic problems for science today. Far and away the largest number of scientific organizations are those devoted to highly specialized disciplines within science; another type of organization, much smaller in number, is concerned with general interests and problems of science as a whole; and there are a few examples of a third type, that concerned specifically with the "social problems" and "social responsibilities" of science. In 1948, for example, there were 208 specialized societies and academies of science attached to the American Association for the Advancement of Science, the most inclusive of the organizations with general scientific interests. These 208 societies were very largely natural science groups, but there were included some social organizations also. This number does not, however, include all the scientific societies that are national in scope.

Another way of appreciating the proliferation of professional scientific organizations is to note those for which a typical scientist is eligible, whether he actually joins them all or not. He could belong to the A.A.A.S., of which we shall say more in a moment as an example of the general type of organization; to the main national society in his field, say, mathematics; to some specialized society dealing with his narrower interest within the field, say, topology; to various state and local branches of the national societies; and possibly to state and local academies of science, some specialized and some cutting across the specialties in science. Most scientists, however, probably belong only to a few of the organizations for which they are eligible, because of the pressure of their professional work and because of limited time and funds.

Now let us look a little more closely at each of these three types of scientific organization. The specialized societies are concerned with the problems, policies, and work of their own disciplines, however specialized. As few as less than a hundred men may be members of a national specialized society. Such organizations are important agencies for providing a loose coordination of the activities in their fields. Their annual meetings are occasions for many valuable informal meetings among their specialist members as well as for the formal presentation of papers reporting scientific research. Sometimes the specialist societies deal with more general problems, for example, the relation of their limited interests to those of science as a whole or to those of the national welfare.

The general societies in science take as their purpose some loose coordination of science as a whole and some concern for the relation of science to the larger society. Two organizations of this kind are Sigma Xi, the national honorary science society, corresponding to Phi Beta Kappa in the humanities and social sciences, and the American Association for the Advancement of Science. It is the ideal of the A.A.A.S. to associate to itself all the specialized national science societies and thus, in some rough fashion, to "stand for" American science. It has already gone some way toward the achievement of this goal. Its annual meetings are vast affairs attended by many thousands of scientists who come to attend the meetings of the several specialized sub-groups of the Association as well as to participate in its more general meetings. The officers of the A.A.A.S. are members of many different specialized societies as well, and their positions require them to take a more general view of American science than they do as specialists. The presidency of the A.A.A.S. is one of the few most distinguished offices to which an American scientist can be elected, and scientists from the universities, from industry, and from Government have all received this high honor, though most presidents have come from the universities. The growth of this general scientific association since its founding in 1848 is a rough measure of the growth of American science as a whole. Note especially the striking acceleration of the growth in the last forty years.

Year	*Membership*
1848	461
1858	962
1868	686
1878	962
1888	1,964
1898	1,729
1908	6,136
1918	9,000 (approx.)
1928	16,328
1938	19,000
1948	42,000[30]

The third type of professional organization in American science is the kind which is concerned for the "social problems" of science, either in regard to the social responsibilities of science as a whole or in regard to some particular issue, like that of atomic energy at the present time. The Federation of Atomic Scientists, officially organized in December, 1945, after informal meetings held during the war at atomic energy research laboratories, is an example of this type of scientific society. Although its general purpose was conceived while the war was still on, the F.A.S. came into active being only when its members felt it necessary to oppose Congressional enactment of the May-Johnson Bill for the control of atomic energy. This bill, scheduled to pass virtually without a hearing, assigned to the military the ultimate control of the new atomic science. In opposition, "a scientists' lobby materialized suddenly in Washington. That lobby was the beginning of the Federation—and, with greater or less effectiveness, it has been in Washington ever since."[31] Although the atomic bomb was thus the immediate occasion for the founding of the organization, the preamble to its constitution expresses an inclusive social purpose: "The Federation of Atomic Scientists is formed to meet the increasingly apparent responsibility of scientists in promoting the welfare of mankind and the achievement of a stable world peace." The more general purpose does not have the same power to attract members as did the more specific one. In the early days of the organization, it had about 3,000 members. In 1950, the national organization had only 1,500 members, grouped into 13 local chapters in 9 states, with members-at-large in 21 additional states. Scientists in American society, like other comparable groups of professional specialists, do not have the time nor the interest to take a very active part in social problems, even those more directly related to science itself. During the past few years the F.A.S. has been less concerned with atomic energy and its control and more with safeguarding the spirit of free inquiry and with promoting "those public policies which will secure the benefits of science to the general welfare."[32] Like many such small voluntary associations, in all fields of activity, the F.A.S. has only one salaried employee in its Washington office. Its work has been chiefly carried out by local volunteers. Nevertheless, the organization has been remarkably successful in creating political pressure for good purposes, not only against the May-Johnson Bill, but on other issues since then.

One thing that holds for all three types of professional science organizations is that they manifest the same pattern of membership participation that voluntary associations in all other fields of interest do.[33] That is, the membership is constituted of a small, active minority and of a much larger, inactive majority. The active minority takes the strongest interest and fills most of the offices. But, unlike some other voluntary associations, the highest offices in science organizations are almost never filled by men who have simply been very active in the group. These highest positions, being symbolic of the status and the values of science, are customarily awarded by vote of the membership as a badge of high professional achievement to the most distinguished scientists in the organization, whether they have participated very actively in its affairs or not. Active participation counts, here as elsewhere, probably, but not nearly so much. We have already seen that election to office as a formal recognition of achievement by autonomous scientific organizations is one of the most important of their several functions.

We have traced out, now, a great change in the social organization of American science. When we consider the proliferation of specialized professional scientific societies, as we just have, we see that science is no longer what Thomas Huxley asserted it to be in 1880, "'a third army, ranged around the banners of physical science . . . somewhat of a guerilla force, composed largely of irregulars." Science is now a standing army of regular professionals, loosely but effectively organized in their own scientific organizations.

Still and all, there do remain a few amateur scientists in contemporary American society.[34] In certain fields of science it is still possible to do some useful work during the leisure hours away from one's regular, full-time job. Some of these fields are astronomy, mineralogy, ornithology, and radio communications.[35] Some amateurs even become expert enough in a special branch of knowledge to be on a par with some of the professionals. But, on the whole, what little amateur work is done is dependent upon the professional work in the same field. The universities, museums, and research institutes serve not only as the source of their knowledge for the amateurs but also as the continuing agencies to whose bulk the small contributions of amateurs may be added. The American Asociation of Variable Star Observers, for example, which has some 130 members in the United States, performs valuable work in its field, but it is valuable only because professional astronomers exist who can organize it and use it in their work. In some fields of science, "pure" physics and chemistry, for example, amateur work is practically impossible because of the degree of training, the amount of time, and the expensive facilities required for satisfactory results. As professionals in other fields, for instance, social work, have discovered, there are great organizational difficulties in using amateurs. Even such an enthusiast for amateur work as Thomas says: "When it comes to the actual use of amateur effort in carrying on important experimentation, or even the gathering of facts to make that experimentation possible, the professional who might consider such a possibility is confounded by the administrative and organizational problems which such a task involves. Use of volunteers in any capacity requires a great deal of planning and supervision."[36] The difficulties that lie in the use of amateurs in science only confirm how important it now is that science as a whole should be in the hands of professionals who have a regularized place in the occupational structure of American society.

In our discussion just above of the functions of professional organizations in science, we mentioned, in passing, the part they play in providing some loose coordination of activities within the specialized disciplines and in American science as a whole. This is an important subject which now deserves our exclusive attention, this subject of coordination and control in American science. We shall return to it again in this book, especially in Chapter Ten where we speak in detail about the social control of science, but for the present it will be enough to pick out certain general features of this aspect of American science.

The most obvious and the most fundamental fact has, of course, to be stated right away, that there is no single, formally recognized, hierarchical organization which coordinates and controls American science. Like science in other "liberal" societies, American *science as a whole* is only informally organized. Science is a pluralistic world in which there is not one but many centers of influence, no one predominant over all the others, although these sub-centers are, as we shall see, related in definite ways. Why this is so, indeed, why it *must* be so, is a matter that goes to the heart of the nature of science. It is a matter that we have already referred to several times in earlier chapters and shall speak of again in later ones; when we come to speak of the social responsibilities of science and planning in science, we shall try to draw the whole problem together. Just now we need to see it as the basic pattern of coordination in American science, the pattern against which the several minor ones take on their significance.

Now just because it is only informally and not formally organized, American science as a whole usually appears to the individual scientist and to the individual layman the way our market economy usually appears to the individual worker or enterpreneur; without any coordination or control whatsoever. But this is not so. However informal and even relatively invisible they may be, there are some definitely structured patterns of coordination running through American science. We have already mentioned that some of these center in the general and specialized professional societies. Another important pattern of coordination and control, both within the several fields of science and between these different fields, is to be found in the informal relations among certain key influential scientists. These are men who are distinguished usually both as scientists and as administrators in science, men of universal prestige and of wide acquaintance. They perform valuable functions in recommending personnel for jobs, in advising on the award of research funds, and, in general, in taking a larger view and a larger responsibility in scientific problems and scientific policies.[37]

Ordinarily this pattern of control through influential scientists is latent, only a few experienced and reflective scientists being aware of its extent and significance. When the purposes of science narrow, however, as they did in the last war, for example, the structure of this informal control becomes somewhat more manifest, although even then it was not apparent to all scientists. In American science, men like J. B. Conant, President of Harvard University, Vannevar Bush, President of the Carnegie Institution of Washington, and Karl T. Compton, then President of the Massachusetts Institute of Technology, men like these and a few handfuls of other scientists wield very large, and beneficial, influence by the way in which they integrate many separate centers of authority and control in science. During the recent war, for example, reports one study of the Office of Scientific Research and Development, the Government agency which was responsible for our scientific war effort, "the administration of O.S.R.D. resolved itself into the triumvirate of Bush, Conant, and Compton."[38] Indeed, Bush himself points out that, although there were approximately 30,000 scientists and engineers working on new weapons and new medicine during the war, there were "roughly thirty-five men in the senior positions" of control.[39]

However absolute the anti-authoritarian values of some of us may be, this is only what is required and desirable for so large and so successful an enterprise. It is only necessary to point out the informal structuring of control in American science because our individual perspectives ordinarily obscure its presence and significance.

Fortunately there has come into print a detailed description of how the structure of informal influence operates, the case being that of the staffing of a war-time science research project.[40] First a scientist was chosen to head the project, chosen presumably at the suggestion of some key figure like Conant, Bush, or Compton. Then, "from his widespread knowledge of his special field, he chose a group of colleagues to assist him." These men suggested the names of another forty to fifty men, who were directly recruited. Most of these men were in academic employment, "and leaves of absence were readily arranged," presumably because of the influence of the men already incorporated into the project. "The recruitment of scientists for the major O.S.R.D. projects," Trytten says in summary, "followed in most cases the 'fanning-out' pattern just described. In the case of perhaps the largest and most successful of the O.S.R.D. laboratories, this process began by a meeting of four internationally known American scientists in a hotel room in New York. From their combined experience they selected the names of forty young and active but completely mature scientists. Through this nucleus the contacts fanned out. . . . The laboratory grew to be large . . . growing to a final level of about one thousand professional scientists and engineers."[41]

The value of such informal coordination of American science should never be underestimated.[42] In his analysis of why the Nazis "failed miserably" in their attempt to make the atom bomb, for which they had much the same opportunity this country had, Vannevar Bush says that an important reason was their poor organization of science.[43] Nor should the great administrative skill of the key American scientists we have mentioned be overlooked. Their successful achievements as administrators dealing with Congressmen, Government officials, and officers of the Armed Forces in wartime Washington have been described as "one of the minor wonders of the war."[44] This success probably would have seemed less remarkable if the similar, though lesser, accomplishments of these men in time of peace were known. Some American scientists have considerable talent as executives, although they prefer to neglect it in order to cultivate their research interests exclusively. The war provided an opportunity to bring out some of this latent talent, as well as that which had already been demonstrated in scientific organizations. For example, the scientific war effort developed a whole new group of scientist-administrators: men like J. R. Oppenheimer, Director of the Los Alamos Laboratory, now Director of the Institute for Advanced Study; Lee DuBridge, Director of the Radiation Laboratory for radar research, now President of the California Institute of Technology; and Frederick L. Hovde, Director of Rocket Research, now President of Purdue University.

We may note one further important fact about this pattern of informal coordination in American science. This is the fact that the central importance of university science research is reflected in the academic connections of most of the key influential scientists. It is also important that some of these academic connections are with the leading institutes of technology, for it is through these research centers, which train many industrial scientists, that the patterns of influence spread out to industrial as well as to academic research groups all over the country. There is, also, through such men as Bush, a link to the privately-endowed scientific research organizations such as the Carnegie Institution, which itself sponsors research as well as subsidizing research by other organizations. In this fashion, informal relations among key scientists join and partially coordinate all the different types of social organization in American science.[45] This kind of informal coordination is an invaluable asset of American science. And not least of all is it important because it is one of the essential conditions of the autonomy of science that such coordination, whether informal or formal, be in the hands of the leaders of science themselves, rather than in the hands of politically appointed non-scientists.

The existence of this pattern of coordination is not something new in science nor is it peculiar to the American scene. In 1864, for example, Thomas Henry Huxley and a group of his friends, eminent scientists all, organized the X Club, a weekly dining society. Huxley later reported that he overheard the following conversation between two scientists who were not members of the club.

"I say, A, do you know anything about the X Club?"

"Oh yes, B, I have heard of it. What do they do?"

"Well, they govern scientific affairs, and really, on the whole they don't do it badly."

This club lasted only for the lifetime of its founders and had considerable influence on appointments and promotions in academic positions and in scientific societies.[46]

Within the general pattern of informal coordination of American science as a whole, two subsidiary patterns of social organization occur. The first is like the general pattern; that is, it is informal. A great many American scientists pursue their research independently as individuals or they carry it on in small "teams" in which the leader is not so much an official administrator as a "primus inter pares," very often in the relation of a master to younger, less experienced apprentices. This informal pattern is indispensable in many areas of scientific research and we shall have to return many times later on to a discussion of its essential functions for the advancement of science. We shall speak of this matter especially in Chapter Nine, where it is related to the social process of invention and discovery. The second pattern is the formal or hierarchical one, the "bureaucratic" one, operating within the general informal pattern, but different from it, and sometimes seemingly in conflict with it. This is a pattern which is increasingly frequent in American scientific research, and it reflects important changes that have occurred in science itself and in its relations to the rest of "liberal" society. "The large research laboratory," says Wilson F. Harwood of the Office of Naval Research, "is a product of the twentieth century." It is especially a product of the last twenty years. "For example, in 1938," he says, "the Naval Research Laboratory in Washington, D. C., the primary center of Navy research and development, housed a few hundred employees in a few buildings and had an annual budget that was substantially under a half million dollars. Today this same laboratory has about one million square feet of laboratory space, about three thousand employees and an annual outlay of $18. million."[47] This is only one of numerous similar cases in university, industrial, and government science in American society.

This change in the social organization of science is, of course, part of a larger movement of change in modern "liberal" society, the change toward an increasing "bureaucratization of the world."[48] When we discuss "planning" in science we shall see how the change in science is affected by the larger trend. We are now concerned, however, with only those conditions which are specifically relevant to the increase of formal organization within science. These conditions include the increased applicability of science, the increased incorporation of science in governmental and industrial bureaucracies, and certain internal changes within science itself.

Let us consider first the significance of increased applicability for the changes that have recently occurred in the social organization of science in America. The great growth of modern science, that is, the vast improvement in its fundamental conceptual schemes, has meant that in an ever larger number of practical situations the theories of science are useful. In such areas as the chemical and electrical industries, in medicine, and in agriculture, to take only a few examples, fundamental science is now applicable. Perhaps the clearest recent instance of the application of fundamental science to a limited, specifiable end is the case of the atom bomb. Now it is in just this type of social situation—where there is a limited, specifiable end—that the formal type of social organization is a most efficient instrument. Ideological objections to "bureaucracy" to the contrary, hierarchical social organization is not a device of the devil. It is rather a great social invention which we are still improving; it is a rational social instrument—with some disadvantages to be sure—for the achievement of limited, specifiable social purposes. And so it has been used not only in our own society at the present time but in other societies at much earlier times. In Classical China, for instance, and in the armies of a great many societies, formal organization has served men well.[49] Little wonder, then, that an increase in the usefulness of science in practical situations has meant an increase in the number of hierarchical organizations seeking to apply fundamental science. Although there are situations in science, as we shall see, where this type of social organization is unsuitable, there are a great many others where the informal type would be equally unsuitable. Here again, we have only to recall the case of the atom bomb.

The suitability of formal organization in achieving a given end, whether in applied science or in certain limited ranges of fundamental science, is especially clear where a social crisis endows the end with unusual urgency. For example, during the recent war, large formal organizations were speedily set up in the fields of electronic and atomic research, the former for the development of such newly specifiable ends of research as radar and proximity fuses. The original plan for war research in science had been to leave each scientist at his own university. That had appeared to be the happiest situation for the scientists themselves. But as the volume of work increased, the need for speedy consultation, for mutual help and instruction, and for frequent intimate contact among scientists working toward the same limited end required that they all be brought together in the same large organization. "The benefits to be derived from teamwork of sizeable groups," says James Phinney Baxter, the historian of the Office of Scientific Research and Development, "were too great to be neglected."[50] Large formal organizations for research were accordingly set up at the University of Illinois, Chicago, Northwestern, M. I. T., and Harvard. On the staff of the Radiation Laboratory at M. I. T. alone there were men from 69 different academic institutions.

The incorporation of science into established bureaucracies devoted primarily to other ends—industrial and governmental bureaucracies, for example—has also increased the amount of formal organization in scientific research. This is in part a consequence, of course, of the increased applicability of science. But even where the applicability of science is not at stake, at least not immediately, there is a pressure upon groups of scientists to organize themselves more formally in order to deal in a regular and orderly fashion with the other parts of the bureaucracy in which they are now incorporated. There is some of this even in universities, we shall see, where scientists must elect "officials" to deal with "the administration." This necessity is all the greater in other types of established bureaucratic organization. Indeed, this pressure toward formal organization exists even where scientists are not actually incorporated into the bureaucracy but depend upon it for financial support even while they remain legally outside it. For example, groups of university scientists have found it highly desirable to set up formal organizations to arrange "contracts" and to deal with the officials of the bureaucracies which supply an increasing share of the funds for their research, namely, the bureaucracies of the Government and the private philanthropic foundations. It is sometimes a condition of the granting of funds that separate research units will be integrated into a larger whole in order to function more effectively *vis a vis* the granting organization.

Certain internal changes in science itself have also contributed to the need for increased formal organization. The first of these is the greater specialization and division of labor in scientific work. Everywhere in social life the division of labor creates the need for coordination and control. Especially in application to specific practical problems must this control, or "organization," in science be formal rather than informal, but this may be true also of research which is more fundamental than applied. The former case, where different scientific specialties must be applied conjointly toward the same practical end, has been remarked on by the late Frank B. Jewett, former Director of the Bell Telephone Laboratories and also formerly President of the A.A.A.S. "In many fields," says Mr. Jewett, "the products will be such as to involve a wide range of physical, chemical, and biological problems so interwoven as to call for scientific attack from many angles, and so we will have large research organizations with specialists and specialized facilities in many fields, all organized to function as a coordinated." He feels that "experience has shown that this is the most powerful, effective, and economical method of handling complex problems. It is greatly superior to any scheme of farming out portions of the problem to individual laboratories. This results from the fact that at all stages of the work the several elements react on one another and that what can be done in one field determines what can or cannot be done in another."[51]

A similar situation requiring the contribution of many specialists in somewhat more fundamental research has been described by Ellice McDonald, Director of the Biochemical Research Foundation. Speaking of a "bacteriostatic and germicidal fraction" obtained from soil bacteria, he says: "It was discovered by the microbiologists and was fractionated by those running the Beams' air-driven centrifuge. It was passed to the bacteriologists and the cytologists to determine its qualities and powers, to the microchemists for analysis, identification and determination of its probable composition, to the organic chemists for fractionation, to the spectroscopists for characterization and for the determination of the spectrographic differences . . . these fractions to the cytologists for toxicity and other experiments on animals, to the microscopists for record of crystalline structure, to the surgeons for study of its external effects upon badly infected wounds, to the physical chemists for the measurement of the physical constants and further study as to its structure. It went as we say 'through the mill.' All these were from our staff work and this is our common procedure."[52] The need for formally organized controls of some kind over such division of labor is obvious. This does not mean that each stage of the work is directed by non-specialists, but still there is a need for the special function of administration.

The second internal change in science which increases the need for formal organization is the change which has occurred in some areas of research, but not all, as a result of the development of instruments which can keep many individual research units busy all at the same time. The most notable case of this kind has occurred in nuclear physics, where the cyclotron and betatron have become indispensable tools of research.[53] Lee DuBridge, now President of the California Institute of Technology, has described the resulting situation. "Several problems," he says, "could be carried on in parallel, and the combined efforts of all groups are needed to keep the machine in operation and to carry on continued improvements."[54] In general, he says, "some of the facilities required for modern work in nuclear physics are so large and so expensive that a large staff is required to operate and make full use of them. . . . I believe it is inevitable that a few great research centers will grow up, and that they will be of greatest importance in the advance of nuclear physics."[55] The need for formal organization in such a situation is obvious.

In sum, then, for the several general reasons we have given, large formal organizations are increasingly common in science. Such organizations can never wholly displace the informal pattern of organization either in science as a whole or in many of its several parts. But even where it is necessary in science, the pattern of bureaucracy disturbs some scientists in just the same way that its establishment in other areas disturbs some of the citizens of "liberal" society. "The imposition of an organizational framework," says L. Kowarski, the Technical Director of the French Atomic Energy Commission, "no matter how flexible or syncopated, may cause a note of regret to professional scientists, most of whom have hitherto lived the lives of independent small-holders."[56]

All apart from this diffuse regret which some scientists may feel, there is room for genuine concern about some of the problems which the pattern of bureaucracy now presents to scientists. Some of these problems of hierarchical organization are inherent in the nature of that type of social organization, whatever the specific purpose of any group so organized, scientific or otherwise. Some of these problems have already been noted in other bureaucracies—in Government, in industry, and in labor unions—and some progress has been made toward understanding them and coping with them.[57] The administrators of scientific organizations may well profit from this general analysis of the difficulties of formal organization. There remain, however, certain problems which are more specific to science. Chief among these is how to preserve the essential autonomy and originality of the workers in "pure" science while guiding them into certain general areas and providing them with assistants and facilities. It is hard to state this formal problem in any more concrete manner, at least in the present state of our knowledge. Perhaps the best we can do is to recognize the necessity to have as the directors of large research organizations men who know from their own intimate experience the nature of scientific research and the problems of its coordination. Only such men can translate the general problem of maintaining autonomy for their subordinates into the specific social situations which the particular scientific specialty and the particular scientific problem at hand require. This may be why some of the most successful large-scale scientific research organizations have been what one successful scientific administrator calls "the shadow of a man."[58] Jewett has spoken, out of his long experience as the director of a very large research organization, of this very problem. "Actually," he says, "what the director and his immediate subordinates do is to provide a proper setup in which men with creative ideas can work freely; to map out the general fields in which progress appears to lie, and finally to weigh the results of research work together with many other factors in deciding how to proceed."[59]

We have seen that this inevitably involves direction in terms of the purpose of the organization, industrial or otherwise, but it also allows the proper scope for the autonomous processes of scientific discovery.[60] Apparently in some cases we can rely on the individual wisdom of particular administrators to maintain the proper conditions for good scientific work. There is a great need for careful empirical investigation of some functioning research organizations to see what makes for success and what prevents the achievement of research goals. "Application of management techniques to large-scale scientific research and development is of such recent date," says one student of these problems, "that we have little documentation of management experience in this area. The outstanding current need is for documentation and evaluation of experience, followed by a codification of findings."[61] The social science of administration, of large scientific organizations or any other, is just coming out of the "common sense" stage.

From the problems of organization and control in American science we turn now to an attempt to indicate the scope and size of American science, that is, the area within which these problems occur. Here we must content ourselves with rough estimates and general impressions, for American science today includes so multifarious and so changing a group of professional activities that it is difficult to describe its size precisely. One basic general fact is clear, however. "American science," says one man whose job it is to keep track of the changing personnel situation in American science, "is riding on a steep up-curve. Everything about it has been on an exponential curve."[62]

Fortunately a recent survey by the President's Scientific Research Board furnishes us some carefully compiled figures which indicate at the very least the order of magnitude of the personnel, funds, and facilities currently available to American science, excluding social science, of which we shall speak in Chapter Eleven.[63] The Board's survey estimates that there were, as of 1946, some 750,000 professional scientists, engineers, and technicians in American society, a group which makes up about ½ of 1% of the total population. Of this number, only 137,000 engage in fundamental scientific research, in technical development, and in teaching. This smaller group is obviously extremely important in American society.

And within even this smaller group there is a nucleus of still more important people. This consists of the 25,000 people who have Ph.D.'s in the physical and biological sciences and who can, even potentially, make contributions to fundamental scientific research. Of the 137,000 total, some 50,000 are estimated to be in the universities and colleges, 57,000 in industrial research, and 30,000 in Government research. From 1946 to 1949 this total increased spectacularly by 30% to the larger total of 180,000.[64] Since 1940 the relative number of scientists in the universities has decreased, that in the Government has remained stable, and that in industry has greatly increased. In 1930 the universities had 49% of the important scientific research personnel; in 1940, only 41%; and in 1947, 36%. This diminution is not, however, necessarily harmful.

This personnel in American science is not considered sufficient. Because it considers science "a major factor in national survival," and because it thinks that "only through research and more research can we provide the basis for an expanding economy," the President's Board recommended in 1946 a policy of planned expansion of scientific personnel over the ten-year period, 1947-1957, without specifying the exact goal to be reached by that later date. We have seen above how rapid has been the post-war expansion, from 137,000 to 180,000 in the numbers of the key scientific group. This is an increase greater than foreseen by the postwar reports on the scientific manpower situation, and indeed it may have been that these reports stimulated part of the great increase. It is unlikely that this rate of increase will be maintained, but large expansion in our scientific personnel is very likely.

We can gauge the scope of American science also by means of the funds it has available for research. Since 1900 the research expenditures of all three types of scientific organization have increased steadily, with Government expenditures growing even more rapidly than those by universities and industry. This pattern of growth has been especially strong since 1930. In that year, it is estimated, American science spent $150. million; in 1940, $350. million; and in 1949, $2. billion.[65] Thus, during the 1930's, the national expenditures on science more than doubled, but the total amount was never greater than ½ of 1% of the total national income. The President's Scientific Research Board made a very strong recommendation in 1947 that funds available to American science reach a minimum of 1% of the national income by 1957. The war, of course, greatly enlarged the amount of money spent for scientific research. During the period, 1941-1945, for example, some $3. billion was spent, most of it for applied research. Eighty-three percent of this vast sum was paid by the United States Government. In 1944 and 1945, more than $800. million was spent in each year, practically all of it by the Government. Despite these large expenditures during the war, the funds spent since have been greater still. In 1947, $1.1. billion was spent; in 1949, $2. billion was spent. The sum will probably go higher, and of course far and away the largest share of it will continue to come from Government funds. Industry spent about one-half of the 1947 total, but the universities, although spending more than they ever used to, absolutely, now spend only 4% of the total. This was a continuation of the relative decline of university expenditures for science, which had been 12% of the total in 1930, 9% in 1940.

Finally, the value of available scientific research facilities is another rough indication of American science. In 1946 the value of research facilities owned and operated by the United States Government alone was $1.5. billion, two-thirds of which was in the Department of Defense. This sum does not include the facilities of the Atomic Energy Commission. Industry had about $1. billion, and all educational institutions, including the universities, had about $300. million in research facilities. Since 1946 research facilities have also increased, in some rough proportion to the expansion of scientific personnel.

Expansion of the sort that has occurred recently in American science and that is likely to continue for a while does not occur, of course, without some planning and without some difficulties. Since personnel is the key element in growth, it is especially important to plan the training program for science, despite the difficulties of this task. It takes some four to ten years to train a scientist, and individual decisions made at one point may not be realistic in terms of the job opportunities that will be available at the end of the training period some four to ten years later. For this reason, "American science, as it continues to grow, must have for its own guidance the best information that can be gathered regarding the trends of its own growth."[66] The collection of this information has for some years now been the responsibility of various groups in the Government science organization, and this function has now been transferred to the National Science Foundation, of which we shall say more later. Such information can help to keep supply and demand for science personnel roughly in balance and prevent any one of three important kinds of imbalance: among the various fields of specialization; among types of skills—administrative, pure science, applied; and among levels of competence, for there is a great need for highly trained and very competent people in science.

Thus, whether measured in the numbers of professional personnel, in the size of financial expenditures, or in the value of its facilities, American science is now an enterprise of considerable scope and of the very greatest national importance. Unless fundamental changes occur in American "liberal" society, the position of science will be maintained, indeed advanced, in the university, in industry, and in the Government.

One last general consideration about the social organization of American science remains, the social characteristics of the scientists who compose the working force in this field. Of these matters, unfortunately, we know all too little, but some things can be said about the class origins of American scientists, their religious backgrounds, and about the place of women in science. More research in this area is an important task for the sociology of science and for immediate practical purposes as well.

Along with other occupational careers in American life, the scientific profession is one in which has been realized the strong American value that talent ought to be rewarded wherever it is found. To what exact extent science has been a channel for upward social mobility, however, we cannot say. Early in this century, 1906, J. M. Cattell made a study of the occupational background of the fathers of the men listed in the first edition of *American Men of Science*.[67] Of his group, 43.1% of the fathers had been in the professional groups; 35.7% in the commercial group; and 21.2% in agriculture. At that time, only 3% of the general population were in the professional group; 34.1% in commerce; and 41.1% in agriculture. Therefore, while the commercial groups furnished about their proportional share of scientists, agriculture produced only about half its share, and the professional groups furnished almost fifteen times their share. It is the latter fact which is most significant, for there seems at that time to have been a very large amount of class self-recruitment, but with some opportunity for social rising. Cattell's "commercial" and "agriculture" categories seem to have included lower and middle class groups, however. The only clear conclusion, therefore, is that a fairly large proportion of scientists came out of the professional groups.

For more recent times, we have another bit of evidence. A Fortune Magazine survey has found that physicist-mathematicians "generally come from middle-class and professional families" and that "a large number of physicists listed in *American Men of Science* are clergymen's sons." Chemists, according to this survey, "generally come from small towns and petit bourgeois parents." And, "the broadest generalization that may be made is that scientists tend to come from the lower-income levels."[68] The only evidence given is that one great source of Ph.D.'s in science is in the smaller, less expensive colleges. There is other evidence for this generalization, though. A recent survey by the Office of Scientific Personnel of the National Research Council has found that "about 90% of graduate students (in sciences) are on some kind of financial support. It shows that graduate students very seldom go to graduate school on their own resources." More people pay their own way in law and medicine than in the sciences.[69]

If we may hazard some generalizations, which ought to be checked and made more specific by carefully designed studies, we might say there is a good deal of social mobility through the scientific profession, but that there is also a great deal of self-recruitment from the professional classes. What social mobility occurs is probably more often from the lower-income segments of middle class than from the very lowest social stratum, but this latter also furnishes occasional recruits for science owing to the great opportunities provided by the system of free public education and scholarships in American society. This general pattern of mobility for science seems, indeed, to be roughly the same as that which holds for the American business elite, according to careful statistical studies recently made in that field.[70] There is a good deal of class and occupational self-recruitment; most of what social mobility does occur is from the lower-income portions of the middle class; and there is some small amount of mobility from the very lowest social class. These are facts which are essential for any recruitment and training program for American science. For example, the National Research Council study mentioned above suggests that about 25% of competent eligible undergraduates do not go on to graduate training because of the lack of opportunities in the form of fellowships. The continued expansion of science requires larger subsidies to talented students who are without financial support from their families.

We know as little about the religious backgrounds and present affiliations of American scientists as we do about their class origins. Recently, however, two Catholic scientists have pointed out that American Catholics do not make their proportionate contribution either to national scientific personnel or to national scientific production.[71] Father Cooper says he "would be loath to have to defend the thesis that 5 per cent or even 3 per cent of the leadership in American science and scholarship is Catholic. Yet we Catholics constitute something like 20 per cent of the total population." The lesser value which Catholics, as against Protestants, place on critical rationality, the more emphasis they put on a teleological conception of the universe as against what we have defined in Chapter Three as "utilitarianism," these are some influences in the Catholic religion and educational system which seem to result in this relatively low participation in American science.

As for American Jews, we have no figures at all, but it is very probable that they are at least proportionately represented in science and learning, at least in the universities. This is what we might expect for two reasons. First, the Jewish values seem to favor learning and empirical rationality. And, second, there is an important relevant factor in the American social system. The Jews have been remarkably socially mobile in the United States, and the free professions, including science, have been more open to them than large industry and many areas of business. Most Jewish scientists would therefore probably be found in the universities and in Government research groups rather than in industrial laboratories, because the industries which support research most heavily have been largely closed to the Jews in all their branches.

Despite their increasing participation in the occupational system generally, women make up another social group which does not play an important part in the scientific profession.[72] In 1947, for example, the A.A.A.S. estimates that only about 1% of its 33,000 members were women. In that year only 2 of the 350 elected members of the National Academy of Sciences were women. Despite all the gains in employment opportunities women had made during the war, in the years just after the war there were still fewer than 15,000 women engaged in all kinds of professional work in the physical, biological, and mathematical sciences and in engineering and architecture. This small figure may be compared with the approximately 500,000 men in the same fields. Even if engineering, which includes two-thirds of the men in these fields, is excluded, women still comprise only 7% of the total in all other fields. The largest professional science specialties for women are chemistry, which employs 42% of all women in science; mathematics, 16%; and 7-8% each in bacteriology, engineering, and physics. Women in science have been limited chiefly to desk and laboratory jobs and have been excluded from jobs requiring field work. There is some prejudice against women as research engineers in industry.[73] Before the war there were even fewer women in science than afterwards. In 1946, for example, the number of women employed in chemistry, some 5000, was three times that so employed before the war.

Women in science suffer career disabilities which derive from the character of their social roles as family members. Women, for instance, are less geographically mobile than men. Women tend to prefer jobs within 25-50 miles of their homes. Studies of the Women's Bureau (U. S. Department of Labor) indicate that the responsibilities of single as well as of married women for financial aid or for personal services to the other members of their families are considerable. Meanwhile, lack of mobility limits the individual's choice of jobs and makes a woman a less desirable employee on jobs where travel or probable transfer may be involved.[74] Many professional women scientists are married.[75] Because they think a woman is much more likely than a man to interrupt her employment for marriage and subsequent family responsibilities, most employers of professional scientists believe men of equivalent experience and training are more desirable as employees. Many married women would, of course, like to work part-time, but most research groups find this to be administratively difficult. Until organizational schemes are worked out for making part-time work available for women, not only in science but in other occupations as well, only a few women will be able to carry on occupational and family roles at the same time.

So much, then, for general considerations about the patterns and problems of social organization in American science. In the next three chapters we take a closer look at these matters by considering in detail the social organization of science in the universities and colleges, in industry and business, and in the Government itself.

VI

The Scientist in the American University and College

ALTHOUGH different kinds of research groups, we have seen, are essential for its advancement, the heart of American science is now in its universities. This was not always so. The university has only recently been so important to science. The early scientists of the modern world, the seventeenth century men who grouped themselves in the amateur societies, we remember, were not members of university communities. Indeed in England, the older universities like Oxford and Cambridge, resisted the growth of science until far into the nineteenth century. Before that time, the happiest place for a career in science was in Government research or in some science institute or museum. The English Geological Survey and the Admiralty were more likely places to find young scientists than the professorships of the universities, which were still objects of political preferment rather than rewards for talent in science. Darwin and Huxley had both served important parts of their training in research expeditions sponsored by the British Government, Huxley as a member of the Royal Navy. In France, and especially in Germany, the university became the stronghold of "pure" science earlier than in England or in the United States; the change had taken place in the first half of the nineteenth century. In the United States, the universities, largely influenced by their German counterparts, not the English ones, did not begin to make their scientific faculties really strong in research until quite late in the nineteenth century.[1]

Not, however, until the twentieth century is the change clear and final. Insofar as science rests on the continual development of conceptual schemes, as we have seen it does in Chapter One, it now very largely rests on the universities in "liberal" society. "The university," said Veblen, "is the only accepted institution of the modern culture on which the quest for knowledge unquestionably devolves. This is the only unquestioned duty incumbent on the university."[2]

That the American scientists who make contributions to "pure" science are now almost wholly located in universities and colleges we can see in the following tabulation made of the listings in the sixth edition of *American Men of Science,* issued in 1938. This edition listed the names of 28,000 scientists, of which 1,556 or 5.6% were chosen by their colleagues as particularly distinguished in their accomplishments and therefore deserving of being "starred" in the listing. The starred men were distributed as follows among different types of research organizations:—

Organization	*No.*	*Per Cent*
Universities and colleges	1,135	73.0
Federal government	128	8.2
Industry and business	131	8.4
Private foundations	120	7.7
State governments	9	.6
Retired	33	2.1[3]

The American university performs two different functions for science. First of all, it integrates science with the rest of American society; and, second, it contributes to the internal development of science by fostering those who make its essential discoveries. The first function is a little less obvious than the second, but equally important. This is because the university is in general a key place for the maintenance, expression, and development of those cultural values of "liberal" society which we have seen, in Chapter Three, are the matrix for a strong science. The university, said Veblen, noting this point, is "a corporation for the cultivation and care of the community's highest aspirations and ideals."[4] The deep approval of science which the university now gives almost without question, the unity of science with the various disciplines of the Humanities in the Arts and Sciences faculty of the university, these represent the implicit recognition of the direct derivation of science from the cultural tradition of "liberal" society, indeed, of its essential place in that tradition. By virtue of its secure position in the university, a position which has been hard-won within the last hundred years, science keeps in close touch with the other scholarly disciplines and with the ultimate cultural values that underlie both kinds of research activity, scientific and non-scientific. More than that, the university trains in all its departments college teachers who pass on the basic values and the cultural tradition of "liberal" American society to the students they leave the university to teach all over the country. Thereby, directly and indirectly, they win their approval for science, an approval which is necessary for the perpetuation of science as a highly respected social activity, and necessary also for the maintenance of financial support to science. Public attitudes toward science do not exist in a social vacuum. In addition to other social agencies like the newspapers, the "liberal arts" universities and colleges are a fundamental support for modern American science. Of course they do more than inculcate values and appropriate attitudes. The university also integrates American science with the rest of its society by providing a generally respected status and career for those who have the ability and the wish to be scientists. We have already seen that the occupational roles of "professor" and "scientist" have a relatively high position in the American hierarchy of occupational prestige.

In addition to fostering the external relations of American science, the university has the second function of promoting its internal development. American universities provide the facilities, taken in the broadest sense to include social atmosphere as well as physical equipment, for the research which underlies the formulation of ever-changing, ever more generalized conceptual schemes. Also they are continually training new researchers, the training usually being in close connection with the current research activities of the mature scientists who make up the faculty. "Teachers who are investigators," says the physiologist W. B. Cannon in his charming and illuminating book, *The Way of an Investigator,* "filled with an ardor for discovery and acquainted with ways to nature's hidden secrets, arouse in young men the qualities they themselves possess."[5]

The training of new scientists involves more than the teaching of theories and techniques, of course. It is also a process of subtle moral indoctrination in the values of science. Thus the university becomes a moral community which not only enforces scientific standards but even incorporates new members into that moral community.[6] For these several reasons, then, the indispensable autonomy of science requires not only a secure place for the university in American society, but also an equally secure place for science within the university.

The essential functions of the university for science are heavily concentrated in American society in a relatively few institutions for the higher learning. In 1939, for example, when there were 90 universities conferring the Ph.D. degree in all fields of scholarship, science included, 3,088 degrees were awarded. Of that number, "more that four-fifths were granted by thirty institutions, more than three-fifths by fifteen, and more than two-fifths by ten leading institutions."[7] Such universities as Harvard, Chicago, Columbia, California, Yale, Michigan, Cornell, Princeton, Wisconsin, Minnesota, Pennsylvania, and Johns Hopkins, and such scientific institutions as the Massachusetts Institute of Technology and the California Institute of Technology are far and away the most important centers for American science. These are the institutions which train the largest numbers of new scientists, have the largest numbers of faculty members listed in *American Men of Science,* and carry on the bulk of the "pure" science research in the United States.

There are, of course, in addition to this key group, other universities and colleges which make contributions to the advancement of science. Just before the recent war, a report of the National Resources Planning Board described the situation for all American universities and colleges as follows.[8] At that time, some ten years ago, there were between ten and twenty academic institutions which were in every sense universities and in which the staffs of *all* departments were chosen primarily on the basis of research ability. In these universities, facilities and time were freely available specifically for research activities by faculty members. In between eighty and one hundred other institutions, research was a recognized part of some departments, but not of all. Some of these departments conducted research at a very high level of competence and at least part of the staff was selected with explicit concern for research ability. In between fifty and one hundred further institutions, some small encouragement was given to research. And, finally, in about twelve hundred remaining American colleges, research was only very slightly or even not at all encouraged. Only his extreme individual initiative could maintain an interested scientific researcher in these twelve hundred colleges. These are the colleges where even those members of the faculty who have won the Ph.D. degree do no research.[9]

Yet because there are some good science researchers in the smaller colleges and also because these colleges train excellent candidates for graduate work at the larger universities, it has often been recommended, only most recently by Vannevar Bush in his report, *Science, The Endless Frontier,* that some of the Government funds for the subsidy of scientific research be allocated to these smaller institutions.[10] Although the bulk of research goes forward in the large universities, the small college should be encouraged to contribute its share to the whole fabric of science. One scientist bases his recommendation that "1 or, at most 2 per cent be apportioned to the Little Researchers" on his observation that although some professors in small colleges do not often contribute research reports to the journals of the national biological and botanical societies, to take but one part of science, much more often their research activities are reported in the proceedings of state and local academies of science. For example, he says, about one-third of the papers in the Illinois State Academy of Science *Proceedings* come from the small colleges of the surrounding region and even from the high schools.[11] An editor of the *Biological Abstracts* has also recently given his impression that there has been "a marked increase in the research production" of biologists in the smaller institutions.[12] It is desirable, and it should be possible, to support the contributions to American science of the men who work where research is not a primary purpose. But the present concentration of scientists and scientific facilities in a relatively few American universities makes it almost certain that they will continue to be the great centers of scientific productivity. Hence their great importance in any program of scientific planning.

In the last chapter we said that much of the internal social organization of science is informally structured and coordinated.

Now we can see that this is true, *par excellence,* of science in the university, and we can look at some of the reasons why this should be so. Ideally, the social structure of the university is in the form of what Talcott Parsons has called "the company of equals" pattern.[13] That is, the university is a social group in which each permanent member of the community of scientists and scholars is roughly equal in authority, self-directing and self-disciplined, pursuing the goal of developing conceptual schemes under the guidance of the scientific morality he has learned from his colleagues and which he shares with them. The sources of purpose and authority are in his own conscience and in his respect for the moral judgments of his peers. If his own conscience is not strong enough, the disapproval of others will control him or will lead to his exclusion from the brotherhood of science. We have seen that the values of science resist external authority. Mature university scientists reject the strict hierarchical pattern of social organization which issues detailed directions and enforces rigid control. In "pure" science in the university, each researcher expects considerable autonomy as his moral right.

Scientific values apart, the "company of equals" pattern in university science is necessary for other reasons. "A university faculty," says one member of such a body in an overstatement for effect, "is composed of people who cannot speak one another's language and who have only the vaguest idea of what one another is doing."[14] That is, in addition to sharing in the cultural tradition of "liberal" society, university scholars each work highly specialized portions of the field of knowledge. This extreme specialization has its virtues, of course, for science often progresses by the union of specialized elements of knowledge that have not been brought together before. A university stimulates this kind of cross-fertilization. But specialization also raises problems, for it makes evaluation and control of the specialist extremely difficult. For instance, it is all but impossible to compare one extreme specialist with another, for the standards of judgment are internal to each of the activities and no third person may be competent to make relative evaluations of the two. Where scientists cultivate the very frontiers of theory, it is difficult to judge the worth of their work and it is dangerous to control it too closely. The difficulty and the danger are old matters in the history of science. The British mathematician, Professor H. Levy, has given the following typical example of the fact that the great majority of scientists are novices in all fields of scientific work but their own. "At the end of 1811," he says, "Fourier submitted his now classic memoir to the Paris Academy on the *Propagation of Heat.* His adjudicators, Laplace, Lagrange, and Legendre, greatest triad of mathematicians of almost any single period, criticized the paper so severely that it was not published by the Academy. As secretary of that institution thirteen years later, Fourier published his results, now become a classic in the *Memoires,* without alteration from the original form. By that gesture he exposed to history the fallibility of scientific criteria."[15]

For these different reasons, in sum, the "company of equals" pattern and informal coordination is required in the university, whether in the departments of science or in the humanities. Each group of scholarly specialists is made up of a number of permanent members, who are usually professors and associate professors, and of a group of aspirants to that status, the assistant professors and the instructors. Those on permanent tenure and those who are "apprentices" make up the self-regulating community in which the several participants are relatively autonomous equals. Invidious comparisons which might be destructive of morale and purpose are thus largely avoided. "For the everyday work of the higher learning, as such," says Veblen, "little of hierarchical gradation, and less of bureaucratic subordination, is needful or serviceable."[16]

The "company of equals" pattern and informal organization are, of course, the ideal. Like other social ideals, this one has its effect upon behavior, but is never fully realized. Probably the nearest approach to full realization of the ideal occurs in those few universities and scientific institutes in which scientific research in the United States is concentrated. Some few of the liberal arts colleges also approximate the ideal. Everywhere, however, and especially in the other universities and colleges, some part of the daily social reality for the faculty is structured along lines of formal control and hierarchical authority.[17] This discrepancy between ideal and reality, which occurs even in the best universities, is not, however, simply the outcome of ineradicable and unexplained human shortcomings. It is, rather, itself the product of those very functions which the university performs for the internal development of science and for its integration with the rest of "liberal" society. As a social organization, that is to say, the university as a whole and each of its several scientific departments require a certain amount of hierarchical organization and formal control as well as a larger amount of informal coordination. In each department some formal authority is necessary to order the relations among the several component specialties and to represent the joint and particular interests of these several specialties in their relations with other departments of the university and with the university taken as a whole. The university administration, for its part, must have the authority to introduce order into the relations among the departments and to advance the interests of the university as a whole with other social organizations with which it has necessary relations. The necessity, then, for these two kinds of social organization in the university—for the "company of equals" pattern and for formally organized authority—require the highest kind of skill in administration if these two are to be blended properly.[18] The ideal president of a university, the ideal chairman of one of its science departments, should be a man who is competent, even distinguished, in *both* technical scientific achievement and in administrative talent. There is no excess of such paragons in America or in any other society.

There are, of course, various ways in which the social organization of the American university, like all social organizations, falls short in the achievement of its ideal purposes. Although the ideal goals of university science are the development of conceptual schemes and the training of new scientists, there are everywhere some deviations from this ideal in the form of system-building for its own sake, cultism, and even careerism.[19] Such deviations are not peculiar to science, to be sure. It is impossible to make even a rough guess of the extent to which these deviations, which arise out of other characteristics and purposes of the university, hamper the development of science. Probably it is fair to say that they are a perennial but pretty well controlled and limited part of university science. Where the level of theoretical development in science is already high and where the moral community of science is strong, there are internal safeguards against too great a subversion of the ideal purposes of science.

One problem of American science which is especially relevant to the universities is the problem of our strength in "pure" science. In recent years, and especially since the war-time destruction of much of European university science, some of our policy-makers for science have expressed the fear that American strength in "pure" science is not great enough. For instance, the President's Scientific Research Board has said that, "as a people, our strength has lain in practical application of scientific principles, rather than in original discoveries. In the past, our country has made less than its proportionate contribution to the progress of basic science . . . we have imported our theory from abroad."[20] Nowadays, of course, the old circumstances of European science and the old conditions of free exchange of ideas no longer prevail. Another scholar has taken the same view as the President's Board. Although pre-eminent in technological invention, says the anthropologist, A. L. Kroeber, "America as yet seems to have produced no men of really first rank, no general initiative and directive leadership, acknowledged as such in Europe."[21]

Although it is true that America has produced no scientific achievements like those of an Einstein or a Pasteur, still there have been many important innovations in "pure" science made by Americans. One might mention here work like that of Joseph Henry on electrical induction; Willard Gibbs in thermodynamics; Michelson-Morley on the speed of light; Millikan on the electrical nature of the electron; Morgan on the gene theory of heredity; Anderson on the positron; Davisson-Germer on wave properties of electrons; Condon on the theory of alpha-particle radioactivity; and Stanley on the nature of the crystalline protein virus. This is of course only a representative and not a complete list of discoveries of a "second order" of significance in "pure" science.[22] It may be, however, no matter how many names could be added to the list, that the United States has been deficient even in people of this second order as well as in "scientific geniuses of fundamental creative significance equal to those of several European nationalities."[23] The Nobel Prizes seem to bear this out. "In some fifty years of Nobel Prize awards in physics, chemistry, and medicine, the United States received only 20, against 119 for Europe and 36 for Germany alone."[24] Lately, however, Americans have been doing better, even apart from those

"Americans" who are refugees from various European countries. Indeed, America may now be "on the verge of the period," says Professor Kroeber, "in which she will bid for leadership in science, possibly seize it."

Whatever the exact nature of this American deficiency in pure science may be, its causes undoubtedly go beyond the responsibility of the university alone. They lie also in the attitudes and the structure of American society as a whole. Professor Shryock has shown how this was so in nineteenth century American society.[25] And yet the university, as the institutionalized center of creativity in pure science, is the locus of the problem, and it is therefore the place at which it can most immediately be treated. One conclusion for an American scientific policy seems inescapable. Science in American universities must be maintained, even strengthened. In part this is a task for the university scientists themselves. But in part it is also the responsibility of the whole society, that is, through its agent, the national government. That this part of the responsibility is understood can be seen in the strong recommendations of the President's Board for increases not only in research subsidies but in scholarship grants to American universities. There seems to be no cause for great alarm that our science will fail us, but we shall be wise to enlarge the sources from which all applications of science ultimately come. The encouragement of pure science in the universities is both morally appropriate and expediently necessary.

The connection of university science with practical applications, and its importance therefor, is nowhere more clearly seen than in the significance of the university for the various scientific professions. Indeed, we may define as a professional scientific occupation one which requires for its successful practice the kind of systematic and generalized theory and knowledge which it is the especial province of the university to cultivate. Such a definition permits a useful distinction to be made among the different types of occupation which apply scientific knowledge of different kinds. It permits us to distinguish, for example, between the doctor who has considerable generalized knowledge of biology and physics and the medical technician who has only limited, specific knowledge; between the "engineer" who has a large understanding of the physical science principles underlying his activities and the "engineer" who has only been trained empirically to make some particular machine or physical structure work. Such a distinction indicates why it is essential that the scientific professions have a close and continuing relationship with the universities, indeed that they be somehow members of the university. Membership in the university community represents for the American professional "schools" a unity in the values of science and also a close relationship with the ever-developing theories which the professions must ultimately apply. All the best medical schools in the United States, for example, are affiliated with universities, and the best engineering schools are either so connected or else form in themselves "partial" universities which carry on large amounts of research in fundamental science. The faculties of scientific professional schools in the universities usually incorporate the two functions; they are at once researchers in pure science and distinguished practitioners in the corresponding applied science. They make considerable contributions to the development of conceptual schemes both in their own research activities and in collaboration with the pure science departments of their institutions. The professional scientific schools of the university are an essential link in the close web of relations between pure and applied science in the United States.

We may now look somewhat more closely at other links in this web of connections between university science and applied science. The relations of the universities, on the one hand, and the Government and industry, on the other hand, are very important in this respect. We have seen that on their side the universities develop new theories, train research scientists, and do some research "to order" for other scientific organizations. On their side, the Government and industry provide some of the tools and even some of the theories for university research. But most important of all, they give the universities heavy financial subsidies in the form of research grants and training scholarships. The changing structure of American science may be partially traced out in the history of these financial relations between the universities, Government, and industry.

Government financial subsidies for university scientific research, which were relatively small in amount in the period before the recent war, increased enormously in the war years, 1941-1946. Moreover, most of the war research was heavily concentrated in the very large universities. Since war urgencies seemed to require that essential research be placed in the large institutions which had the personnel and facilities, very little thought was given to the possible harmful effects of this situation on the smaller universities and colleges.[26] This pattern of support for university science has persisted into the post-war period, a survey by Benjamin Fine, of *The New York Times,* shows.[27] First of all, as to the size of Government subsidies. In the academic year 1949-1950, the Federal Government gave about 200 academic institutions more than $100. million for research. This represents an estimated increase in amount over what was given in the pre-war years of no less than 500%. As a result of this increase, many academic institutions now have the bulk of their scientific research directly subsidized by Government funds, the largest shares of which come from the Department of Defense, the Atomic Energy Commission, and the Department of Agriculture, but other Government agencies contribute some small amounts as well. The distribution among types of universities is the same as it was during the war. "For the most part, the Federal money is concentrated in the larger universities and the big-name technological institutions," says Mr. Fine. We shall consider in a moment some possible harmful consequences of these facts. Although there were hundreds of subsidized research projects, in literally every field of science, more than one-third of all Federal funds in 1949-1950 went to the engineering sciences, and another half of these funds went to the physical and medical sciences. More important, almost all of these funds were for relatively applied research. However, in contrast to some other Government agencies, the Office of Naval Research made a special effort to subsidize worthwhile basic research. In 1949-1950 the O.N.R. supported some 1200 projects in about 200 institutions, the total cost being $20. million. Nearly 3,000 mature scientists and 2,500 graduate students worked on these 1200 projects.

The ever greater predominance of Government funds in university science has not been accepted with complete satisfaction by university scientists and administrators. "Many educators interviewed by this writer," says Mr. Fine, "were worried lest the colleges and universities 'slant' their research too heavily in the direction of applied projects at the expense of fundamental research. Other officials feared that the highly concentrated research program will 'freeze out' the smaller colleges and harm the whole field of research." This draining off of scientific talent into relatively applied work, even in the universities, may be one source of the apparent American weakness in fundamental science. With respect to another important aspect of scientific organization, however, there seemed to be less to fear from Government funds. University scientists who spoke to Mr. Fine did not complain of Government interference in their internal administration. "Primarily," they said, "the Government is interested in results, and does not interfere with the college administration." Some might still ask, though, autonomy for what?

University science has also been receiving financial support for a long time now from American industry. This support is of two kinds. First, there are unrestricted gifts, grants, and graduate fellowships which are given without any expectation of direct return to the donor company or industry. These are considered contributions toward the general development of science as a whole or of some special field of science. Secondly, some funds are given for quite specific research projects of immediate benefit to the sponsors. Probably the first kind of support is absolutely much smaller than the second, though it is not necessarily any less important relatively to university science. In pure science a little money may often go a long way. Neither kind of support from industry is new, but the volume of both types has increased greatly in the most recent years. For example, 302 companies reported to the National Research Council in 1946 that they were supporting research in colleges and universities by means of approximately 1800 fellowships, scholarships, and research grants. By contrast, in the year 1929, only 56 companies had reported only 95 similar awards. Another indication of the financial dependence of university science on industry: in the 1946 edition of the National Research Council's directory of industrial research laboratories there is a list of about 300 educational institutions which do some research for industry. State universities, land-grant colleges, and technological institutes comprise the largest part of this list.[28] The large private universities are not so immediately dependent upon industrial subsidies as are these latter institutions, which are constrained partly by political and partly by financial reasons to have this dependence.

University scientists have sometimes been as disturbed by this subsidization of their research by industry as they are by Government subventions. And at least one leading industrial scientist has joined his university colleagues in expressing concern for the possible bad effects of the increasing subsidy from industry. "Universities at the present time," says C. E. K. Mees, Director of the Eastman Kodak Research Laboratories, "are tending more and more to embark upon industrial research in cooperation with industry, much of this so-called research being really development work of a type calling for energy and inventive ability rather than for scientific imagination. This is likely to be far more disastrous to the free spirit of inquiry in the university than the receipt of support from such an organization as the National Science Research Foundation."[29] Of the Foundation we shall say more below.

Thus in both cases, in the case of Government support and in the case of industrial support, there is some fear that the increasing financial dependence of university science on outside organizations may have harmful consequences for its autonomy and its productivity in pure science. It would be alarmist to think that these consequences are inevitable, but of course they are more likely to occur if inadvertence prevails among university scientists. Government, industry and the university will have to join hands in forestalling the excessive use of scientific resources on applied research, a circumstance which could ultimately do more harm than good to applied research itself.

One matter arising out of their increased dependence on external funds which American university scientists have had to pay more attention to recently is the problem of patenting scientific innovations. We have seen in Chapter Four that it is an ideal of the scientist interested in "pure" research that his discoveries should not become his private property but rather the common property of the community of scientific peers. We have seen also that it is functionally necessary that this should be so, for otherwise parts of scientific theory would be removed from the public domain and thereby the advancement of science would be hindered. For these reasons the university scientist is opposed to patenting his discoveries. A report on *The Protection by Patents of Scientific Discoveries,* published in 1934 by the Committee on Patents, Copyrights, and Trademarks of the American Association for the Advancement of Science, summed up the following chief sentiments and reasons against patenting of research which university scientists have:—

"1. That it is unethical for scientists or professors to patent the results of their work:

"2. That patenting will involve scientists in commercial pursuits and leave them little time for research;

"3. That publication or dedication to the public is sufficient to give the public the results of the work of scientists;

"4. That patenting leads to secrecy;

"5. That a patent policy will lead to a debasement of research;

"6. That patents will place unfortunate stricture on other men who subsequently do fundamentally important work in the same field;

"7. That it is debatable whether one man should receive credit for the final result he obtains after a long series of studies has been carried out by others before him; and,

"8. That the policy of obtaining patents will lead to ill feeling and jealousy among investigators."[30]

The existence of this early A.A.A.S. report indicates that the problem of patenting is not a wholly new one for university science. Even when American university science was relatively more devoted to pure research than it has been recently, patentable discoveries sometimes occurred as unexpected results of such research. Some scientific innovations in the university always had immediate commercial application or required control in the public interest. In such cases, almost always the problem was solved by assigning patent rights to non-profit organizations set up for the purpose of managing the patent and dealing with it commercially. In recent years, because of the very great increase in research done by the universities for the Government and for industry, the problem of patenting has arisen much more frequently. As a result, universities and colleges everywhere have had to re-examine their old policies in regard to patenting and many of them have drawn up new formal statements of policy in the last five years. The general problem having come to the attention of the National Research Council, it sponsored a study, right after the war, which indicates that the universities have not yet settled the matter and are pursuing "a wide diversity of practice . . . even at the same institution. There is no common pattern of policy statement, administrative procedure, recognition of the inventor, determination of equities, assignment requirement, patent management plan, distribution of proceeds, or protection of the public interest."[31] In accord with scientific values, however, "at most institutions the compulsory asignment of patent rights is not considered desirable, except when necessary in connection with co-operative or sponsored research."[32] University scientists have complied with these values by assigning the rights to any patented discoveries either to local corporations specifically set up for that purpose or to national organizations like the Research Corporation, a non-profit patent management foundation which is handling patents for an increasing number of universities and colleges. By transferring patents to an "ethical" agency of this kind outside the university itself, scientists who make patentable discoveries can avoid the harmful consequences which all of them so much fear.

In its turn, the Research Corporation has been able to help university science by distributing the income on the patents it holds for the advancement of pure research. From 1912, when it was founded by Frederick G. Cottrell, who gave it the patent on his electrical precipitation process, still its most valuable property, until 1945, the Research Corporation had given $1,250,000. to university research. Among the research projects to which it has given assistance are the cyclotron, the Van de Graaf high voltage generator, the utilization of solar energy, computing machines, and the synthesis of vitamin B1. After the war, a special five-year program of grants-in-aid to fundamental research was endowed with $2,500,000. by the directors of the Corporation. In the first year, 163 grants amounting to $865,000. were given to scientists in 32 different states, with preference being given to young men who had been in war research and who were going back to academic institutions.[33] Thus is the "disinterestedness" of science realized; thus is its value of "communality" achieved.

University scientists would, of course, prefer to do only such research for agencies of the Government or for industrial enterprises as would allow them freedom of patent assignment. But many of them now face the moral dilemma that they cannot have both the outside funds and this freedom. This is only one aspect, to be sure, of the larger problem which university science now faces, the problem of maintaining its autonomy for the development of conceptual schemes while taking funds and doing research for Government and industry. On the finding of solutions for this problem the future achievements of American science as a whole now depend.

In another area, that of its relations with the private foundations, American university science has had a somewhat more favorable experience with this problem of autonomy. These non-profit organizations devoted, as the charter of the recently-established Ford Foundation puts it, to "scientific, educational and charitable purposes, all for the public welfare," have found it easier than government or industry to square their own goals with those of the universities. Beginning roughly in 1900 and continuing to 1920, the foundations, especially the largest ones like the Rockefeller and Carnegie Foundations, gave most of their money for general educational purposes. Even by 1920, though, they were giving $2. million a year for basic scientific research. Since 1920, scientific research has been an increasingly important part of the program of the foundations. From 1921 to 1930, $22,677,544. was expended for the natural sciences alone; in the 1930's, more than $30. million was spent.[34] The emphasis on basic science has been intensified in the 1940's, and 1950's.

The foundations have tried to aid university science by giving as much of their funds as possible on science's own terms. Toward this end, lump sums have been given to various university research councils, which in turn have distributed the money among their colleagues in the fashion which seemed most useful to themselves as scientists. In addition, the foundations have granted subsidies directly to many individual research projects, and among these they have especially favored pioneering work. One of the most important uses to which foundation money has been put is graduate and postgraduate training fellowships. The best known of these fellowships in science itself are those awarded by the National Research Council with funds granted by the Rockefeller and Carnegie Foundations. Former National Research Council Fellows make up a good part of the present elite of American science. The foundations have been able to give their money wisely for research grants and fellowships because so many university-trained scientists serve in the administration of the foundations and on their grant- and fellowship-committees. The most fruitful expenditure of funds for scientific research requires a large measure of active participation by scientists themselves. It is significant that the Government's National Science Foundation has been given the title of "foundation" and not "authority," and that its administration is largely in the hands of scientists.

VII

The Scientist in American Industry and Business

THE HISTORY of the development of American industry to its present very high level of productivity and efficiency is in considerable part the history of the development of American science and technology. We cannot stop here to trace out this history, much of which is still to be written. We need only to note the present result of that history for the social organization of science within American industry. We need only to note, that is, the basic fact that the widespread and intensive use of science is now an essential condition of the successful functioning of American industrial enterprise.

This condition of its success is not something which American industry leaves to chance. Science has been taken out of the garrets and workshops of "genius" inventors and incorporated into the very heart of the industrial firm. This has come about because the achievements of scientific research laboratories have now become one of the indispensable components of the most important decisions of industrial policymakers. As a result, in most large industrial organizations, and especially in the newer ones which rest most closely of all on scientific discoveries, the Director of Research is more than a scientist and more than an administrator of other scientists. He is now customarily also a vice president of the company and a member of its top planning group. He has become the executive who must take a larger view of the purposes and conditions of his company in the light of the general potentialities of science. His functions have been indicated in the remark once made by a member of the Board of Directors of the General Motors Company about Charles F. Kettering, G.M.'s famous research director and inventor. Mr. Kettering, it was said, is "a cross between a goad and a soothsayer."[1] Of course industrial executives take more things into account than the recommendations of the Research Director when they make large decisions, but they seem to be increasingly aware that he speaks for scientific actualities and scientific possibilities which they ignore at the peril of their success. There is, says a report of the National Resources Planning Board, "a widening acceptance of the thesis that research promotes the growth and increases the earning power of companies."[2] Indeed, American industrial leaders now explicitly show their appreciation of the value of applied science by presenting the advantages of research in their annual reports to stockholders and in their prospectuses for new stock issues. Strong science research is now good business.

As a consequence of this favorable attitude toward the use of science, even before the recent war American industry employed more than 70,000 research workers in more than 3,480 laboratories at an estimated annual cost of $300. million. By 1947 the total of expenditure had risen to $500. million, probably with a corresponding increase in personnel. We are here speaking only of the uses industry makes of natural science. Our figures would be somewhat larger if we included the various uses it makes also of social science. Of these we shall speak in Chapter Eleven.[3]

But if American industrial research organizations now probably lead the world in size and in quality too, this has not always been the case. Among the industrial nations of the modern world, Germany was the first to make systematic use of highly-trained scientists, men with Ph.D. degrees, in its industrial enterprises. In the 1860's and the 1870's, before Great Britain and the United States had begun to do so, the German chemical industry, for instance, employed university-trained scientists to control and advance its new techniques of manufacturing synthetic dyes as a by-product of coal-tar. This particular German priority was all the more remarkable since it had been an Englishman, W. H. Perkin, who had been the first to synthesize an aniline dye. During the last quarter of the nineteenth century, American industry turned for assistance to university professors and commercial research chemists only sporadically, without an understanding of the new powers of science. This is not to say that science was not important to the development of American industry at this time. It was rather the lone scientist and the independent inventor who made during this period the discoveries which became the basis for new industries. These scattered, unorganized researchers, often empirical "boil and stir" men, were financed by industrial capitalists only after their discoveries had been made and tested. Among the men of this kind, now great heroes of early American industry, were Thomas A. Edison, who invented the phonograph, the incandescent light bulb, and a great many other things; John Wesley Hyatt, who in 1872 began the manufacture of celluloid, the first modern plastic substance; E. G. Acheson, who discovered the process for making abrasive carborundum and lubricating graphite; and, Charles M. Hall, the discoverer of the electrolytic process for producing aluminum metal from its ore.

Not until early in the twentieth century did American industrial firms begin on any considerable scale to organize their own research departments and to employ university-trained scientists. The first industries to do so were those which had themselves been born in the laboratory, like the electrical industry. Older industries were much less quick to bring science into their activities. "At the beginning of the century," says Frank B. Jewett, retired Director of the Bell Telephone Laboratories and himself one of the pioneers in industrial research, "the first timid adventures were being made with a handful of young men lured from the ranks of teaching."[4] Jewett has commented on how much more favorable to industrial science university scientists themselves have become over the last forty years. If industry was slow to take science in, the universities were then not anxious to have their young men leave teaching and research for the new industrial laboratories. The new pattern for the use of science spread slowly in this country until the first World War, when the actual and potential uses of science for industry were demonstrated for everyone to see. Thereafter, the number of organized research departments in industrial organizations increased very rapidly, from about 300 in 1920 to 3,480 in 1940. Over the same period, the personnel employed in scientific research in industry has increased from approximately 9,300 to more than 70,000. There has also been an increase in the number of large research organizations, of which we shall say more later. Only 15 companies had research staffs of more than 50 persons in 1921; in 1939 there were 120 such companies.

Not all the industrial workers classified as "research" personnel are professional scientists, of course. Some are technical workers and others are maintenance workers. The ratio of professional to technical to maintenance personnel in industrial research organizations is something of the order of 2:1:1. The following table indicates the distribution of personnel in 1940.

Occupational Classification of Industrial Research Personnel

Type of personnel	*Number*	*Per Cent*
Professionally trained:		
Chemists	15,700	22.4
Physicists	2,030	2.9
Engineers	14,980	21.4
Metallurgists	1,955	2.8
Biologists and Bacteriologists	979	1.4
Other professional	909	1.3
TOTAL PROFESSIONAL	36,553	52.2
Other technical	16,400	23.4
Administrative, clerical, maintenance	17,080	24.4
	70,033	100.0[5]

It will be noted that chemists and engineers make up about three-fourths of all the professional people, with a very small representation from the biological sciences. This is because the largest industrial users of scientific research workers have been the chemical, petroleum, and electrical industries, which have depended very heavily in their origins and for their advance on scientific discovery. These are the fields in which the largest numbers of trained engineers exist. Indeed, it is the expansion of "scientific industry" that is chiefly responsible for the great increase that has occurred in the number of engineers in American society. In 1880, for example, there was one member of a major engineering society for every 30,900 of the United States population. By 1900, there was one for each 8,900 people; by 1920, one for each 2,120; and, in 1949, one for each 910 people.[6]

Although industrial research is sponsored by companies of all sizes, including the smallest, "the bulk of industrial research contributions are being supported by a rather limited number of large corporations."[7] We have here the same pattern of concentration that exists in university science. Only the very largest corporations, as we shall see further in a moment, can afford a large research staff. Consider the relation between the average size of the research staff and the financial size of the sponsoring corporation shown in the following figures:

Tangible net worth of company	*Average research staff*
1 million	13
10 million	38
100 million	170
1000 million	1,250[9]

Financial size is not, however, the sole determinant of the location and size of industrial research organizations. Another important factor is management policy, that is, the attitude of management toward the usefulness of scientific research. As a result, organized research laboratories are found in all industrial areas in the United States and in practically all types of industry.

American industrial research organizations themselves, the actual "laboratories," range in size all the way from something like the Bell Telephone Laboratories, "by far the largest research laboratory in the world, employing over five thousand people and costing about $3,000,000. a year," to the single engineer which is all that some small companies can afford.[9] The very largest research organizations, which we shall consider again further on, can do all kinds of scientific work—pure research, applied, and so-called "development" work. The small companies, which of course make up the great majority of all industrial enterprises, have research staffs which can only handle relatively simple applied and development problems. For other kinds of scientific research which they may need, they must turn to outside research organizations. Fortunately, there is a large number of different kinds of such research facilities now available to the smaller companies. They can turn to commercial research laboratories, like the Arthur D. Little Co., for example, one of the oldest of the proprietary research organizations in this country. Such laboratories are also used by the larger companies for some kinds of research, for instance, when special skills or apparatus are required which are not available in the larger company or where the problem is one which does not require close operating contact with the factory. Small companies can turn for help also to the colleges and universities, to certain non-profit research laboratories, like the Armour Institute, and most important of all, to their own trade association research organizations.

American industrial technology is now so much the stronger because the trade association, whether in its own laboratory or not, is an important sponsor of research for its component, small-sized members.[10] When the trade association does not have its own research organization, it can use the facilities of the commercial research laboratories, or it can get its research done through fellowships and research grants to educational institutions. We have seen in the last chapter that some 300 such institutions are making themselves available for this kind of research. In addition, the National Bureau of Standards in the United States Government does a great deal of research for trade associations as well as for individual companies. Altogether this makes a wide variety of scientific resources available for the trade associations and, through them, to the smaller companies. The trade associations sometimes also offer other scientific services than research itself. For instance, some of them review all the domestic and foreign publications that have any possible application in their industry and advise their members of new discoveries and developments it is to their interest to know.

The scientific services of the trade associations are not, however, the whole solution to the needs of smaller industrial firms. Very often the trade associations find it difficult to finance research of sufficient size and duration. Some member firms lose interest when immediate results are not forthcoming, and they withdraw financial support. It is important that trade association research, when it is carried on, be so managed that it benefits all members equally and not merely the special concerns of a clique of firms within the association. Some appropriate types of research work for trade associations are the improvement of standard products for the industry and the development of new outlets for these standard products. These types of research are subsidized, for example, by the National Canners Association and the Paint and Varnish Association. These and other trade associations have found it best to have a special research committee to direct the research program. A committee of this kind, composed of members of the association, can mediate between the industry and the research organization, stating the industry's problems to the latter and carrying its solutions back to their colleagues.

Another important outside scientific research asset for American industry are the privately endowed but non-profit research institutes like the Mellon Institute, which is connected with the University of Pittsburgh, the Battelle Memorial Institute, and the Armour Research Foundation. At these several institutes, industrial firms of any size can have research projects done from which they alone get the results. However, at the same time, they can have all the advantages of having the work done in a very large research organization, with its many different projects all going on at the same time, its excellent library and general equipment, and with its regular administrative officers and directing scientists. Some idea of the number and kinds of research projects carried on may be had from the following sample account of activities for one year at the Mellon Institute. "In 1944-45, there were 94 industrial research programs in operation, employing 242 scientists and 232 assistants. The service staff of the Institute numbered 169, and total expenditure was slightly more than $2. million. The subjects under investigation were diversified: for instance, catalysis as related to the synthesis of butadiene; utilization of corn products, such as starch, oil, and zein; improvement in waste disposal in streams; structural glass; coal and coke products; synthetic lubricants; properties of cotton fibers; petroleum products; organic silicon resins; and industrial hygiene."[11]

Another way of enlarging the research facilities available to American industry is through collaboration among several research organizations in the smaller and middle-sized companies. Most of such collaboration has up to now consisted of exchange of technical information, but recently a new joint activity has been added. This is the exchange of experience on the organization and administration of research in industry. A new collaborative venture, the Industrial Research Institute, affiliated with the National Research Council, has as its purpose the improvement of management methods in industrial science. For, as we have already said in Chapter Five, there is still all too "little information or experience available on how to organize and manage research. A research organization has peculiar characteristics of function, operation, and personnel that do not easily lend themselves to customary business management practices."[12] Acting upon this deficiency, the Industrial Research Institute seeks primarily to help the middle-sized research organizations, those with fewer than 100 people on their staffs. The very largest industrial research organizations have by now very skillfully arranged their activities in this respect. The Institute will hold periodic meetings for the informal discussion of such common problems as organization, personnel management, project selection, budgeting and accounting, selling research, university relations, and patent procedure.

Now we may finally come back to the very large industrial research organizations. These have been the most strikingly successful, perhaps, of all American industrial research groups, the ones which have reaped the richest rewards for their sponsors. The numerous personnel of these large laboratories carry on elaborate programs of research in which many different kinds of scientific work are being maintained simultaneously and in close interconnection. These organizations represent the vanguard of industrial science. As we have seen, research centers of this kind are a product of only the last thirty to forty years. Willis R. Whitney began research for General Electric in 1900, but for some time he had few colleagues. It was in 1902 that Charles L. Reese organized the Eastern Laboratory, the pioneer among the several research activities which the DuPont Company now maintains. Frank B. Jewett, whom we spoke of earlier as the Director of the Bell Telephone Laboratories and who is now retired from that position after long service, began his telephone research in 1904. Among the first of the large industrial laboratories was the Eastman Kodak research organization, established under C. E. K. Mees in 1913. In 1917 Westinghouse Electric first set up organized research as a separate activity. Other industrial giants which now have very large research organizations are the Dow Chemical Company, United States Rubber, and Standard Oil of California.

We have also seen that only corporations of great wealth like those we have just mentioned can afford large research organizations. Good industrial research of any kind, and especially relatively fundamental research, is expensive for a number of reasons.[13] In the first place, good research talent and good research facilities are not cheap. The scientists and the equipment available in the very largest industrial research organizations are the equal sometimes of the very best that can be found anywhere, even those in the universities. Indeed, some university researchers can be attracted to industry by the superior facilities which companies are willing to provide for research activities in which they are interested. This seems to have been the case in the transfer of Dr. Wallace Carothers from Harvard University to Dupont to work on what eventually became nylon fiber. In the second place, research is expensive because there may be a period of anywhere from five to ten years between the original fundamental scientific conception or "hunch" and its application in an actual industrial process or product. Fundamental research must therefore be subsidized by large sums of money for several years without return. We have seen that Dupont was willing and able to wait nearly ten years for Dr. Carothers' research on high polymers to pay off in the manufacture of nylon. "The Badische Anilin-und-Soda Fabrik," says one student of industrial research, "spent fifteen years of patient research and five million dollars in 'patient money' before they learned how to make synthetic indigo."[14] And, last of all, the developmental work which lies between pure research and industrial application is also very costly, not only in equipment but in engineering talent. Only the wealthiest company, sometimes, can afford to build the expensive pilot plant which is usually necessary to test the industrial practicability of scientific discoveries.[15]

Perhaps we should add the risk element to the reasons why fundamental industrial research is so expensive. It is true that the results of research are often highly remunerative, but it is also true that research achievements can never be guaranteed. "Industrial research is an adventure," C. E. K. Mees has said, "it is even a gamble, although one in which the odds are on success."[16] The financial risk of this irreducible element of "gambling" is one which can best be absorbed by companies with very large capital. Moreover, although some of the results of research are quite obviously profitable, some of its products, says the report of one very large research organization, that of the Standard Oil Company of California, "are not easily measured."[17] In this obscure case, the large corporation can more easily take it on faith that research is profitable, without inquiring too closely into every product of its research organization. The less wealthy company does not often have the financial margin for "faith" of this kind. "The fact is," says one research director, "that measured in terms of the usual business standards, research is an expensive luxury."[18] A wrong hunch, he says, may cost a quarter of a million dollars. Because research has this speculative quality in the short run, there is a tendency in industrial research for boards of directors to pour in money in good times and to cut it off in bad times. During the "dark days" of 1931, for instance, Bichowsky reports, the order came to the research departments to "cut out work on all projects not producing profit."

Even in good times, of course, large financial outlays for research are not something which even the very wealthiest corporations make irresponsibly. Recently a group of corporations having very large research organizations, which they consider permanent parts of their companies, compared their "costing" experiences in this field and found that they were still seeking satisfactory answers to some common problems. Among these problems were "1. A suitable principle or formula for the overall provision of research funds, both current and long-range; 2. Procedures for the control of expenditures . . . to assure the selection of the most promising projects; 3. Cost control, through which funds may be diverted into more productive channels of research and development, or saved; and, 4. The evaluation of results."[19] Despite the difficulty of making cost estimates and keeping cost accounts in industrial research, still they are necessary instruments of planning and control. For although there may be some greater margin for risk in the large than in the small company, nevertheless industrial research must be made to pay on some roughly demonstrable basis, whether it is financed in the long run or the short run.

We have already spoken of this matter of the profitability of industrial research and now perhaps, since it is so essential among the conditions under which industrial research organizations operate, we may offer some further testimony on this point. Both our observers have had long and intimate experience in this field. The first is Mr. R. E. Wilson, a leading executive in the oil industry. "Why did our industry," asks Mr. Wilson, "which 30 years ago employed less than 40 research workers . . . expand this activity until today it employs several thousand full-time research workers and would like to hire as many more over the next few years?

"I should like to be able to tell you that the oil industry's research was due to a broad, public-spirited interest in the welfare of the country. Frankness, however, compels me to let you in on a little secret. . . . The real reason why our industry has spent hundreds of millions of dollars in research and development during the past quarter century is that we thought we could make a profit by so doing!"[20]

The second is Mr. Bichowsky, industrial research director, from whom we have already heard. "To work on soap films," he says, "just for the fun of it or to increase the sum of human knowledge is not good business or good sense except under certain conditions.

"To justify pure research requires the same reasoning that is required to judge the most impure research. The difference is simply in the factor of time. Langmuir played with poorly evacuated bulbs because he or Whitney (in this case both) knew that new methods and new ideas were needed if lamps were to be improved."[21] General Electric, which employed Whitney and Langmuir, is in business for profit like everyone else. Its industrial research must eventually "pay off," however long the run in comparison with some smaller company. It hardly need be said that this industrial imperative of profitability does not deny the significance of other industrial purposes, like those of turning out a good product, or gaining the respect of one's manufacturing colleagues and of the general public.

Even on the understanding that research will be profitable over the long run, it is not easy to establish and maintain successful large-scale industrial research organizations.[22] They must be administered to a large extent like "bureaucracies," and this, as we have seen in Chapter Five, presents some special problems when a scientific group is involved as well as the general problems attendant upon all hierarchical organizations.[23] In industry, as in the university or in the Government, the Director is an especially important person in the large-scale research organization. He must be a man of great administrative ability in addition to being a scientist of some achievement and experience. The Director of a large industrial research group must be particularly aware of the different patterns of social organization, formal and informal, and of the different types of leadership, which are suitable for relatively pure and relatively applied scientific work. In such an organization, probably most of the work is applied, work in which researchers can be assigned fairly specific tasks by hierarchical superiors. The more fundamental research, however, requires more self-direction for the working scientist and his team of assistants and staff. In the latter case, the good Director serves to define the general area of research, as Whitney of General Electric did for Langmuir, and is skillful enough to grant just the right amount of independence. Skill of this kind is best learned through actual experience in industrial research. Fortunately, the men who now direct American industrial research have actually "risen through the ranks" and have acquired their great skill through long experience.

In the Bell Telephone Laboratories, for instance, which "have no large number of professional scientists pursuing their own inquiries in an independent and largely unco-ordinated manner," the Director allows "a certain latitude for the more original and individualistic effort."[24] One mathematician was allowed to work at home and come to the laboratory only a few days a week. "The physicist, Davisson, who won the Nobel Prize for research on electron diffraction, has pursued lines of inquiry of his own choice much wider than the field of telephone communications," though they lie close to the interests of the telephone company. "However, the majority of personnel of the Laboratories are working on assigned projects under general direction. The degree of freedom of inquiry achieved by Davisson and the 'mathematician who worked primarily at home' is the exception."[25] We could use more knowledge of this kind. Unfortunately, despite their very great social importance, we have no detailed, systematic study of how any one of these large industrial research organizations really operates. We have studies of their formal organization patterns, but not of the informal, day-to-day problems of recruitment, career patterns, incentives, and relations between scientific specialists and general executives. Most of our knowledge has been gained from casual accounts of these organizations given incidentally to other purposes.[26]

The contributions of American industrial science not only to the larger national welfare but to the immediate advancement of the conceptual schemes on which science as a whole rests are now considerable. American industry and the rest of American society as well, therefore, have a great obligation to maintain and strengthen industrial science.

VIII

The Scientist in The American Government

GOVERNMENT science may seem a new thing under the sun to many of us, but actually the scientist has had a place in the American Government almost since the very beginning of our national history. Our Government has never been loath to employ scientists who could give it assistance in its problems, whether of war or peace, and so science has always been as much connected with the Government as with the university and industry. The history of Government science, still as much unwritten as that of industrial science, is an essential part of the history of American science as a whole.

For instance, one of the longest continuous records of Government scientific research has been written by the Department of Agriculture and its predecessors. This record, which goes back well over one hundred years to the early part of the nineteenth century, was at first in the hands of a very few individuals. When the Department of Agriculture was first established, it set up its scientific program by hiring a single chemist, a single botanist, and a single entomologist. In the years since then, each of these single individuals has expanded into whole bureaus of scientific workers in all the problems of agriculture. And now, after a long evolution, says T. Swann Harding, historian of the Department, "the Agricultural Research Center alone, where a staff of over 2,000 persons works . . . is almost certainly the largest agricultural research institution in the world."[1] And this is only a part of the Department's scientific resources.

This pattern of growth has been the pattern for all scientific research in the United States Government. At first, in the nineteenth and sometimes into the early twentieth centuries, only a few scientists were employed in any given department. Over the last fifty years, however, the numbers of Government scientists have increased manyfold. By the time of the years just after World War II, the Government was the home of some 30,000 specialists in the physical, biological, agricultural, and engineering sciences. About one-third of these are agricultural scientists and another third are engineers of various kinds working for the Department of Defense.[2] In 1947, $625. million were spent by Government departments for physical and biological research by these 30,000 scientists.

Although there are certain types of concentration of scientific effort in Government science, as we shall see, the full range of research done by Government scientists is very wide, indeed probably universal in its scope. "Virtually every scientific discipline and sub-discipline," says the President's Board, "is explored in the Federal Government's program."[3] The very largest part of Government research, as we might expect, is applied, accounting in 1947 for $570. million out of the total of $625. million spent in that year. But some relatively fundamental research is also carried on. Always, however, as is the case also in American industry, this more fundamental research is approved in the conviction that it will, in the not too long run, be useful for some practical purpose. Judging from the work of the Department of Agriculture alone, the achievements of Government science in both pure and applied fields have been of tremendous value to the American people "both in scientific advancement and in monetary terms."[4]

Before the recent war, the largest part of the Federal research budget was expended for agricultural research. In 1936-37, for example, when the Government spent $120. million on research, (about 2% of its total budget for that year) one-third of this amount went to agricultural research, one-fifth to military research, and the rest for other purposes.[5] Since the war, the balance has shifted quite definitely to research for military purposes, and the military departments of the Government now pay for five-sixths of the total Federal research program.[6] The Government has always, of course, spent very much more on research in the natural sciences than in the social sciences. Just before World War II, three-quarters of the regular budget was spent on natural science research. In the period of the Depression, during the '30's, however, in addition to the one-quarter of the regular budget, most of the emergency funds for research were given over to the social sciences.[7] Although it is hard to obtain precise figures for the distribution, this predominance of the natural sciences certainly holds for the post-war Government research program as well.

Since Government research, like that in industry, is under a compulsion to demonstrate its value in the shorter run, it has been concentrated in military and in agricultural research activities for fairly obvious reasons. Military research is clearly the Government's own vital concern, and agricultural research has become its concern so heavily because no single farmer has ever been able, as have some individual industrialists, to subsidize scientific research. As a relatively under-privileged socio-economic group in American society throughout much of our history, the farmers have sought for and succeeded in getting a great deal of aid from the Government through its program of agricultural research. We often forget that we have been through an agricultural revolution in the last hundred years as well as an industrial revolution. In this agricultural revolution Government science has had a large and pioneering part to play. "For example," says one study of the matter, "no other research was done in the field of nitrate fertilizer until the Bureau of Chemistry and Soils had given 10 years to the work, during which period costs were reduced by two-thirds. When the profitable methods had been developed, the industry established nitrate laboratories, hired the Bureau's nitrate specialists at three to five times their Government salaries, and set about commercializing the process."[8]

Nevertheless, despite the great social contributions made by Government scientific research, its public prestige has never been very high, probably because the prestige of all Government employment in the United States has not been very high. There are very few American scientists who consider Government employment the most satisfactory type of career. Even among scientists already employed by the Government, one survey of attitudes showed, only 37% felt that the greatest career satisfaction could be had in the Government. Of the university scientists, only 1% felt this way, and only 5% of industrial scientists shared their feelings. No wonder it has sometimes been hard to recruit scientists for Government research. "Of all groups combined, only 11% preferred a Government career," says the survey summary. "31% preferred industry and 48% the university environment. The remaining 10% preferred consulting work or some other activity."[9]

Nevertheless, whatever the prestige their work had, many very competent scientists have worked and are still working for the Government. Like industrial research, Government science varies quite widely in its average quality. Probably Government research has been best where it has not been in direct competition with industrial research and with the high salaries which industry can afford to pay. Government research has been particularly good in at least two areas, those of agriculture and medicine. In the latter field, for example, the Government research program constitutes one-quarter of the total national effort. The Government is doing all kinds of research in medicine and it has considerable influence on the rest that is done in the universities and in industry. Of the several Government agencies doing research in this field, the Public Health Service spends perhaps the largest proportion of its funds for basic research; the Army and Navy medical work is more in the applied direction. Altogether, in 1947, the Government employed about 3,500 medical research personnel, who were using excellent facilities, had adequate funds, and were given considerable freedom in the choice of problems. As a result of favorable conditions like these, "the Federal agencies engaged in medical research," says the President's Board, "have maintained a high quality of work, respected throughout the world."[10] Contrary to some popular prejudice on this score, Government scientific work is not necessarily mediocre. It must always be kept in mind that "there is as wide a difference in atmosphere and morale among various laboratories within Government as there is among industrial establishments and universities."[11] There is no necessarily fixed inferior quality in all Government science, and its present variability suggests that where it is now weak it may be capable of improvement. This, at least, has been the conclusion drawn by both of the large investigations recently made of Government science, one in the '30's by the National Resources Planning Board and one after the war by the President's Scientific Research Board

All other factors apart, one important element in the problems of Government science is the organizational one. Government research organizations have labored under two kinds of organizational handicaps, and it will be worth our while to look at each of these in some detail, and in this fashion we can see some of the ways in which the average quality of Government science can be improved. The first organizational difficulty is that our Government research departments are in many respects controlled by general rules which apply to all Government organizations, non-scientific and otherwise, and which are not suitable in important degree to the successful functioning of a scientific research group. And, secondly, Government research has suffered as much and more than industrial or university science from those organizational problems which afflict all "bureaucracies." Neither of these two kinds of difficulties is in principle insoluble, and we shall look at some efforts that are already being made to mitigate them in Government science.

Almost all Government scientific workers, like most other Government employees of course, come under the general regulations of the Civil Service. Now for all their virtues, these Civil Service regulations have up to recently, partly because of their formal nature and partly because of the way they have been administered, been more suitable for routine and mediocre jobs than for positions requiring specialists of high competence. As a result of these rules, for instance, the Government has been at a disadvantage in the competition with colleges and industry for the recruitment of scientists. The universities and colleges, in this connection, have the advantage of personal acquaintance with applicants for jobs, and most such positions go by personal recommendations based on intimate knowledge of the candidates. For its part, industry annually sends out teams of energetic talent seekers to communicate in person to college undergraduates the great attractions of jobs in industrial science. In contrast, the man who has wanted a Government science position has had to search out his own opportunities, take a written examination that may occur at some place to which he must travel at his own expense, and then wait some more time still before a Government job was even offered to him. "The competition," says a man who had considerable experience in hiring Government scientists during the war, "is between a dynamic process of energetically seeking out the best available personnel and hiring with a minimum of delay or formality and a passive process aimed not at seeking good candidates but at resisting unworthy ones."[12] In this situation, Government science has every year lost many potential employees to industry. But it is possible to speed up the Government hiring process, and happily the Civil Service has recently made changes in its rules which accomplish just this. Something has also been done by Government science organizations and by the Civil Service Commission to develop a program of information for college students and young scientists about the advantages of Government work: its good facilities, good starting salaries, and genuine opportunities for career advancement. For instance, the National Bureau of Standards, in order to compete with the "rushing" of the best college seniors by industry, "rushed" the best juniors for summer jobs and thereby they achieved Civil Service status. Very often these juniors wanted to come back after getting their college degrees, and they found this easy to do because of the Civil Service status they already had.[13] But more general improvement in the recruitment process is necessary if the Government is to hire more of the better young candidates for jobs in American science.

Government-wide Civil Service regulations also cover the salaries that are paid to Government scientists. On the whole, the salaries these rules allow are higher on the average than they are in the universities but lower than in industry. For the younger men, however, Government salaries are just as good as they are even in industry. But the highest salaries paid under Civil Service, those around $10,000, are much lower than the highest ones in industry. As a result of this salary ceiling, it is difficult for Government research to retain its best men after they have served for several years and reached the maximum salary. It is not impossible to keep them, of course, because some men are attracted by other features of Government science work, such as particularly excellent research facilities in their field. But by and large, there is a drain of talent out of Government research, and thus the Government becomes a kind of training school for industrial scientists. This is of course an indirect benefit to the whole national community, but these advantages need to be weighed against the direct loss to Government science itself. It has been recommended that the top salaries in Government work for specialists of many different kinds, not just scientific specialists, be increased somewhat, say to $15,000. It is probable that an increase of this kind would help to retain many men at all levels who are now lost to industry. Such an increase is unlikely, however, except as part of a general change in salary levels throughout Government work.

Government scientists suffer other disadvantages under Civil Service rules. Because they have been devised primarily for administrative purposes other than scientific ones, these rules tend to create more opportunities for promotions and top salaries for men with seniority and administrative talent than to men with specialized technical competence. Government scientists, therefore, must sometimes choose between job advancement and scientific opportunity. Since Civil Service general policy does allow for high job ratings to professionals without administrative duties, this dilemma could be made less common by creating more high-ranking positions open to full-time research personnel who want job advancement without administrative responsibility. Administrators are essential, of course, and some scientists do make excellent full-time or part-time administrators, but there are some men who are anxious to get ahead and yet unwilling to leave research work. The Government has need of scientists of this kind.

Scientists require other kinds of special treatment which are not customarily granted Government employees under the Civil Service rules but which are necessary for good scientific work. For instance, Government research organizations should be able to establish in-service training programs for their personnel; they should be able to allow them to go to scientific meetings without loss of pay and preferably even with some subsidy of their expenses; and they should be able to grant leaves of absence to men who want to take further university or other kinds of scientific training. Industrial research organizations, especially the large ones, provide all these opportunities for their employees. Since they are important to good scientific work and indispensable means of career advancement, the lack of these privileges makes Government scientific work much less attractive than it would otherwise be. In these respects too, fortunately, there has recently been improvement. Indeed, there has even been some precedent for the changes that are being made. Even before the recent war, the Department of Agriculture supported an extensive program of in-service training, with some fifty courses being given to 3,000 students by excellent teachers. Since the war this kind of program has been extended to all Government scientific departments. After making a survey of the graduate training needs of about 10,000 Government scientists in the Washington, D. C., area alone, the Civil Service Commission's Advisory Committee on Scientific Personnel has arranged excellent local training opportunities. "Local educational institutions," at its behest, "have set up fully-accredited, off-campus graduate science courses for the convenience of Government scientists in the Washington area. More than 700 physicists, engineers, chemists and mathematicians have enrolled in one or more of 26 such courses, usually offered immediately after working hours in a room close to their place of work."[14]

In other local areas, and in particular scientific laboratories, similar facilities are now offered to Government scientists. For instance, in the Navy Electronics Laboratory in southern California, there exists a program of educational opportunity which is even more elaborate than the Washington one. In addition to formal courses at both undergraduate and graduate levels offered in cooperation with the University of California at Los Angeles, there are informal discussion groups for the whole Laboratory which cut across the several specialized work groups. There are also intensive seminars for specialized scientific subjects. But "the most interesting and unique part of the program is that filled by the visiting consultant. Leading scientists of known and outstanding ability in fields related to, or directly in line with, the work of the Laboratory are brought in for limited periods to act as consultants and directors."[15] Such a consultant may organize a research program to be carried on after he has himself left, he may give advice on work under way, or he may for a time participate directly in such work. He may also give occasional lectures and seminars. Even though it is hard to measure, his influence on the ideas and outlook of the Laboratory scientists is beneficial for their increased potential ability as well as for their immediate work. This program indicates what is possible in Government science, says the author of our report, who has come to feel that "there is no reason why, eventually, the point cannot be reached at which a stimulating atmosphere of learning and free discussion exists throughout the Laboratory to the same degree as at academic institutions."[16]

These, then, are some of the situations that require special rules for the scientists in Government employ or that require the careful administration of general rules already in existence. Much has already been done and still more can be done, even in a system of Government regulations which is necessarily designed to eliminate political interference and other forms of discrimination, to create the appropriate organizational conditions for Government science. These changes will make Government employment a more attractive career to scientists and will help to improve the average quality of Government scientific research.

But there is a second kind of organizational difficulty which hinders Government science, the kind of difficulty which, we have seen, occurs in hierarchically structured research organizations in industry and the university as well. We refer to the conflict between certain conditions of scientific autonomy and certain administrative necessities of the formal type of social organization. In respect of this class of problems, too, improvements are being made in Government research organizations which will make them generally more effective instruments of the national welfare.

Let us, for example, consider some of the problems of financing and planning Government scientific research. Like the administrators of other research organizations, Government science administrators are required to plan their research budgets some time ahead and to account for expenditures that are actually made. Both of these duties they find somewhat easier to fulfill in the case of applied than of relatively fundamental research. Now all Government budget appropriations, the science ones included, must be stated on a year-to-year basis and must be quite specifically described and justified. But a great deal of scientific research, even of the applied kind, is very hard to describe in detail before it is under way and some of it is even harder to plan on a single-year basis. Of course scientists in Government must understand the difficulties which administrators have in drawing up research budgets which can be approved by their organizational superiors and by Congress itself. But it is desirable to have at least some Government research that could be budgeted on a three-year or a five-year basis, in order to provide the essential continuity that is required especially in basic research. It is often hard to employ scientists, and it is certainly wasteful to build facilities for their use, unless relatively long-term financial commitments can be made. It is also desirable to have some Government research appropriations made in lump sums for broad programs instead of in many small amounts for many specific projects. The present Government budgetary and accounting procedures, so far as science is concerned, says the President's Board, are "costly in terms of money and in the time and energy of administrators and scientists."[17] On the whole, Congress is only willing to grant money for fairly specific researches, and it has made very few basic grants of authority for any and all research which a Government science organization may consider necessary. Such a basic grant has, however, recently been given to the Office of Naval Research, and this practice could be extended, with due care, to other executive agencies. Congress faces a dilemma here, for although it is increasingly aware of the importance of long-term, fundamental research, it is also seldom willing to surrender the control it has over executive agencies through the power of appropriations. Insofar as Government budget appropriations for science continue to be on the fiscal year schedule, the administrators of its research organizations will often be compelled to act in accord with the laws of Congress and not in accord with the needs of scientific research.

Perhaps the greatest conflict at the present time between administrative necessity and scientific autonomy in Government science is over the increasingly extensive requirement of secrecy in research.[18] Many scientists are unwilling even to work for the Government because of this requirement, which offends their values, yet in a great deal of current Government research, particularly for defense, secrecy is necessary, or at least it is thought to be by Congress. The kernel of the difficulty lies in deciding what work must be secret and what need not be. On the one hand, administrative caution presses toward making the secrecy restrictions stronger and more inclusive. And on the other, scientific values predispose toward rejecting the secrecy regulations altogether. Somewhere between these two institutional imperatives the line of secrecy must be drawn. It will always be hard to draw the line, because it is always hard to weigh precisely the relative gains for science and the relative gains for national defense security. The difficulty can only be diminished by mutual understanding between administrators and scientists of the different pressure which their values and their roles impose on them. If administrators must learn to give as full scope as possible for the anti-secrecy values of the scientists, then these in turn must learn to acknowledge the burden of responsibility for security which falls on the administrators.

Another matter in which administrative requirements and scientific ones differ is that of the evaluation of the work which is done in Government science organizations. Of course scientific work must undergo something of the same kind of administrative and legislative review that all Government work is subject to. Usually this review is cursory and accepts without question the scientific work that has been done. But sometimes other than scientific standards of evaluation are applied, say, the standard of acceptability to Congress, much to the distress of the working Government scientist. Sometimes also sheerly administrative standards of evaluation are vague. In reply to the question, "How is research work of the unit evaluated?", addressed by the President's Board to the several Government science agencies, these organizations attempted to give satisfactory answers, but, says the Board, "the statements are singularly vague and unilluminating."[19] Some research agencies now use advisory committees to evaluate work, committees which are competent to consider standards of scientific ability as well as the more easily applied administrative criterion of immediate utility. While there is an inevitable pressure on Government science to take demonstrated utility as the standard of evaluation, there is no reason why the criterion of the advancement of fundamental science cannot also be applied. We have seen that in all science, industrial as much as Government, too much emphasis on immediate practicability is likely to yield smaller returns in the long run. Government science administrators still have much to do in convincing Congress of the value of fundamental research.

One agency which has confronted these several difficulties of administering a Government science organization and has tried to work out various solutions for them is the Office of Naval Research. Its experience should be helpful to other Government science agencies.[20] For instance, the O.N.R. uses distinguished non-Government scientists as advisory consultants to those who will actually carry out its research programs. The consultant is used exclusively for scientific problems, not administrative ones, and he is used only when and where he is needed. This device permits a fruitful use of part-time, civilian scientists in Government work. The O.N.R. has also sought to maximize the usefulness of its full-time employees by reducing their non-scientific duties to a minimum. For example, where it is absolutely necessary that the scientist also be an administrator, special administrative assistants are employed to perform the most routine of his non-scientific tasks. As far as possible, indeed, the Naval research program has tried to reduce even the low-level scientific work of mature scientists. Such work is given over to aides and laboratory assistants. Adequate clerical and statistical aid is provided. In short, the aim has been to organize the Navy's research laboratories so that each scientist is working at his own full competence as much of the time as possible. This is difficult to do, of course, because a laboratory is not like a production line operated by semi-skilled labor which is easily trained, exchanged, and replaced. One way to achieve maximum use of actual competence would be to broaden the scope of personnel placement beyond the single laboratory, perhaps even beyond a single agency like the O.N.R. Then scientists could be moved about among jobs, among laboratories, and among agencies to the place where their skills were most profitably employed for the Government and for themselves as well. There are limitations, of course, on flexibility of this order, but it might be practical in some measure for Government science as a whole.

This suggestion for change indicates that some of the solutions for further integrating administrative necessity with scientific autonomy in Government science lie in greater coordination among the multiple Government scientific agencies. Some of this coordination already occurs, not only through informal liaison and exchange of information but through formal inter-agency committees. Also, scientists do move about among Government organizations in search of better opportunity. But a formally-established supervisory agency with limited general responsibility for all Government science seems desirable. This might be one of the functions of the National Science Foundation, whose other functions, for civilian science, we shall consider below. The improvements for Government science which might come out of more coordination of its several parts should not be exaggerated, however. The number of Government science organizations is now so great, the amount of science they do so large, that there are definite limits on efficient coordination. Still in this area of Government activity, as in so many others, as the Hoover Commission reports have shown, there is room for improvement.

There is another problem of coordination within the Government, a problem which reaches beyond science but concerns it very much. This is the problem of the relations between Government scientists and the top-level military planners, relations which apparently are still not as satisfactory as they might be.[21] During the recent war, scientist-administrators saw that there were many cases in which only they could anticipate things that science might develop for the uses of national defense. As the need existed then, so it still exists for the incorporation of scientists into the military groups which carry on the highest level of strategic planning.[22] Of course the responsibility for ultimate decision remains with the military and foreign policy planning groups, and even, in the case of the atom and hydrogen bombs, with the President himself. It is not the role of the scientist to make such decisions. But, increasingly, the knowledge and the fore-knowledge that the scientist has is an essential component of these strategic decisions. Any program for the greater coordination of Government science as a whole must therefore make some provision for its effective integration with some of the larger reaches of national political decision.

In all that we have said up to now about Government science we have omitted the work done by the Atomic Energy Commission, but perhaps we ought to say something about it even though to a great extent it is cloaked in secrecy by Congressional statute. Although we know few of the details, we do know that peacetime Government science has entered a new phase of size and significance with the establishment of the A.E.C. Born almost directly in pure science, atomic energy was developed into maturity during the war with more speed probably than any other major discovery in man's history. And the very intensive wartime program, although now somewhat diminished, has its continuation in what is still a vast scientific and engineering effort.[23] The purposes of the Atomic Energy Commission have been stated as follows by Professor Smyth, one of its members: "The first is to make more and better weapons. The second is to develop possible peacetime uses of atomic energy; and the third is to develop such scientific strength in the country as is needed in the long run to support the other two."[24]

These are goals which obviously require the use of both pure and applied science, and accordingly the Commission has established a number of National Laboratories in which the different kinds of necessary scientific work can be most effectively carried out. There are five of these laboratories, each an organization of an unknown but admittedly large number of scientists: Berkeley (California), Los Alamos (Southwest), Oak Ridge (Southeast), Argonne (Chicago), and Brookhaven (Northeast). Berkeley was the only one of these which existed in some form before the war, when it was by far the largest American university center for research in nuclear physics. Its purpose was little changed during the war, only pursued on a much larger scale, and since the war it has been almost wholly subsidized by the A.E.C. Despite this, Commission control is relatively weak, Berkeley having been left largely in the hands of the university scientists who first established it. As a result, mostly basic research is still carried on there. By contrast, Los Alamos Laboratory was established during the war and is still used for the specific applied purpose of research on the atomic bomb itself. The Argonne Laboratory, closely connected with the universities which exist in the Chicago area—Chicago, Northwestern, Illinois, and Indiana—, was during the war and remains a Reactor laboratory carrying on chiefly pure research. The Oak Ridge Laboratory, also a product of the war, has mixed functions, being both a service and a research laboratory. Lastly, there is the Brookhaven Laboratory near New York. Brookhaven was established just after the war because of the presence of a large number of scientists in the universities of the northeastern part of the country, men who could be used on a part-time basis for the program of almost wholly basic research at Brookhaven. No one university could afford to build the research facilities which this laboratory has, yet these facilities were needed if the universities were to be able to carry on certain kinds of physics research relevant to atomic energy. The rotating staff of Brookhaven Laboratory brings university and Government science into close and continuous relations.

The difficulties that attend the administration of scientific research organizations so large as these A.E.C. National Laboratories are obviously great.[25] Fortunately, the members of the Commission, some of whom are experienced scientists, know something of the problems both of science itself and of large research organizations. They know, says one of their members, the different conditions under which pure and applied science flourish.[26] They know the dangers of "big" science: getting too large, red tape, and not getting new blood. Indeed, they are even aware that they may succeed too well and draw off more scientific talent from the universities than they should. The future of science in the United States now depends a good deal on the success of the A.E.C. National Laboratories, not only with respect to their scientific discoveries but also on their skill in devising effective organizational arrangements for the integration of university and Government scientists.

The Atomic Energy Commission is not the only Government agency using outside research facilities. We have already seen that American industry receives research assistance from the universities and the Government. Just so, many Government research agencies are using the universities and industry. Indeed ,all United States Government scientific agencies now support research in various non-federal organizations with grants of money, contracts, and graduate fellowships.[27] Most of the research now done for the Government by outside scientific organizations is, like that done by the Government agencies themselves, military research. For example, nearly 80% of all the papers presented at the 1948 spring meetings of the American Physical Society came out of projects supported by Office of Naval Research funds.[28] Among Government departments, the older ones with research agencies, like the Departments of Agriculture, Interior, and Commerce, have been using outside aid much less frequently than the military. Indeed, during the war itself, these Departments actually helped out in military research.[29]

Research relations between the Government and university and industrial organizations have presented special problems. One of these has been the development of standardized procedures for making grants and contracts so that the interests of both parties to an agreement could be preserved. On the whole, where it is more difficult to tell how a research will turn out, as in basic science in the university, the Government has made general grants of funds. Contracts with more specific terms have been feasible for applied research. Even with this flexibility, Government budgetary procedures have sometimes imposed on outside research agencies the same planning difficulties that occur in the Government's own scientific organizations. Many research projects cannot be planned and executed in less than from three to five years. But Federal grants, like Federal appropriations, are limited to one year. Some partial change would be as desirable here as it would be within the Government itself.

In the making of grants and contracts, one administrative matter is of particular importance. That is, now that the Government is expending such large sums in grants to outside agencies, it needs to hire the best administrator-scientists available to supervise the making of these grants. The private philanthropic foundations which grant subsidies for scientific research have long since learned that it is hard to spend money wisely. They have accordingly chosen highly competent men as administrators of their funds and paid them good salaries. In the Government too, the highest quality personnel is necessary for the administration of the grant program.

We have already mentioned another problem coming out of these new research relations between the Government and the universities and indusrty. We have seen that the number of patentable inventories occurring in Government-sponsored research has increased greatly as a result of the increased use of outside facilities. This has raised the question of a proper patent policy for the Government, as it has also raised it for the universities. The patent problem is not a new one for the Government any more than for the universities, but it has a new urgency. "The question of what disposition of patent rights from Federal research will best serve the public interest," says one discussion, "is one that has been pondered and mooted in all branches of the Government during the past fifty years."[30] Numerous Government studies of patent policy have been made over that period, the latest being one completed just after the war by the Department of Justice. "On the basis of that study, the Attorney General submitted a comprehensive report to the President in May, 1947, recommending as basic policy that all technology financed with Federal funds should be owned or controlled by the Government."[31] But the objections from industrial firms doing research for the Government to this policy were very strong and it has never been adopted. Industrial research directors feel that such a policy would make it impossible for them to do any research and development work for the Government. As a result, Government agreements with industry and universities have been variable and *ad hoc* with respect to the disposition of patent rights. Unless there is a major change in the general character of the relations between Government and industry in the United States, it seems most unlikely that the policy recommended by the Attorney General will be legislated by Congress.

We come, finally, to the last aspect of Government science we need consider. Even before the end of the recent war, two facts had impressed themselves on those responsible for the use of science by the American Government. There was first the fact of the tremendous importance of science to the well-being of American society in peacetime and in wartime alike. And second there was the fact that only the Government can now furnish sufficient funds to maintain American science at its present high level of quality and productiveness. The diminution of funds for university research was especially likely to cramp the general advance of science. No longer are gifts from private wealth and returns from capital investments sources of sufficient funds for university science, especially in the face of inflationary costs. Although money of this kind still has a great, perhaps crucial importance for the universities, nevertheless they cannot carry on as the country needs them to carry on without some kind of Government subsidy.

Recognition of these two facts has made an increasing number of Government officials and leading scientists feel that the Government ought to take on a new function, namely, responsibility for some direction and coordination of American science as a whole. The recommendations of the men who feel this way have been most explicitly stated in the wide program of "planning" and development proposed by the President's Scientific Research Board and by Vannevar Bush's report, *Science, The Endless Frontier,* both of which we have already referred to several times. These are charters for the new social organization of science, at least as some Americans see its possibilities. Among the recommendations made by the President's Board are items like these: the United States must enlarge its expenditures for science as a whole and increase the number of trained scientists; in the future we must place a heavier emphasis on pure as against applied research, the former being relatively weaker; toward this end the Government should increasingly subsidize pure research in the universities; the Government should also have a large number of scholarships for undergraduate and graduate science students, the holders to choose their own field of specialization; the Government should appoint a committee for the coordination of research among the Government agencies, the committee to include civilian as well as Government scientists; and, last of all, the Government should set up an organization, to be called the National Science Foundation, to supervise these several new scientific activities.

The different parts of this general program have been realized at different times since it was first proposed just after the recent war. For example, increased Government funds for university science have been available all along, but the establishment of the National Science Foundation, as the coordinating agency for the whole national science program, was not legislated into actual existence until 1950. There were a number of reasons for this, but an important one was the conflict that arose over Government's enlarged functions for American science. During the long delay in the establishment of the N. S. F., the scientists of the country themselves seem to have been overwhelmingly in favor of it. For instance, in reply to a Fortune Poll question, "Do you favor the creation of a National Science Foundation with federal funds to stimulate basic and applied research?", 90% of academic scientists, 91% of Government scientists, and 81% of industrial scientists said "yes."[32] Some leading scientists, however, particularly from the universities and industry, strongly opposed any such coordinating agency unless it was basically in the control of scientists who were outside the Government itself and not ultimately responsible to the President.[33] The view of these men found considerable support not only in business groups but in Congress as well. The President, however, opposed it as contrary to established Government procedure and to good executive practice. Indeed, acting upon these principles, he went so far as to veto a bill in 1948 which set up a Foundation with a Governing Board which was not responsible to him. This was only one of several attempts in the years 1945-1950 to frame legislation which would be acceptable to Congress, the scientists, industrial groups, and the President.[34]

In 1950, finally, as we have said, a bill establishing the National Science Foundation was passed by Congress and approved by the President. The bill is pretty much an enactment of the recommendations of the President's Board that we have already listed. It states as one of its primary purposes "the promotion of basic research and education in the sciences." This purpose is to be realized by aid to universities and by subsidies to research projects and for scholarships. The Foundation will not itself operate any research laboratories but will confine itself to making grants of money and planning the development of American science. The Foundation is expected to "appraise the impact of research" on the general welfare of the United States. It has many specific functions defined by its general purposes: it is to evaluate the research programs of the other Federal Government agencies; encourage the exchange of scientific information among scientists of the United States and foreign countries; keep a roster of scientific personnel; initiate and support specific researches for the military when requested to do so by the Secretary of Defense; and establish special commissions to survey special fields of science and recommend general programs for those fields.

These activities of the Foundation are to be in the hands of a Board of twenty-four members appointed by the President with the consent of the Senate. Board members are to be chosen for six-year terms from among persons, says the bill, "eminent in the fields of the basic sciences, medical science, engineering, agriculture, education or public affairs." The bill advises the President to consider nominations for these posts from the National Academy of Sciences, associations of universities, and other scientific and educational organizations, seeking in this way to give legal status to the influence of autonomous scientific groups. The President is also to appoint a Director with the consent of the Senate. The Director will receive $15,000. a year for a six-year term. Important decisions, however, says the bill, are to be made by the Board of Directors itself. This is an awkward administrative arrangement still, but probably some satisfactory procedures for getting the business of the Foundation done will be worked out. The bill recommends four divisions within the Foundation: medical research; mathematical, physical, and engineering sciences; biological sciences; and scientific personnel and education. The social sciences are not included, but neither is there any explicit prohibition on support of social science research and education, as there was in some earlier versions of the bill. The bill recommended an appropriation of $500,000. for the first year, when the Foundation would not do much beyond organizing itself. Thereafter a limitation of $15. million a year was imposed on appropriations directly for the Foundation, although it may receive additional funds from other Federal agencies.[35]

In accordance with this bill, shortly after its enactment the President nominated a Board of Directors, and legislation was introduced into Congress for the first-year appropriation of $500,000. for the Foundation's expenses. But what followed showed that not all opposition to the Foundation was dead. First the appropriation was cut down to $250,000., and then even this smaller amount was not voted. For the second year, the President requested the maximum amount allowed by law, $15. million, but Congress voted only two per cent of this amount, $300,000. On this budget the Foundation could do little more than keep its organizational apparatus in existence. For the third year, the President requested Congress to appropriate $14. million, but again Congress cut heavily, though this time only to $3.5. million. The Foundation will, therefore, at last be able to realize part of its program. About half of its funds for the fiscal year 1952 will be expended in grants for basic research in biology, medicine, mathematics, the physical sciences, and engineering. The other half will be used for 400 fellowships for graduate students in science, the fellows to range from first-year graduate students to post-doctoral trainees. These first fellows will begin their subsidized studies in September, 1952. Even this brief history indicates that absolute opposition to the National Science Foundation cannot long have its way. The need for basic science and for the coordinating functions of the Foundation is too great; good sense and urgent necessity will prevail before very long. Government science, like Government in general, seems to have a relatively more important part to play in American society in the future.

IX

The Social Process of Invention and Discovery: The Role of the Individual and Society in Scientific Discovery

IN THE FIRST volume of *Das Kapital*, Karl Marx observed that men usually pay so much attention to the actual physical product of their labors that they are blinded to the social relationships and social processes out of which these physical products come. This habit Marx calls the "fetishism of commodities," and it has been noted since Marx's time that it is a habit which occurs in many areas of social life. One of these places is the one with which we have now to deal, the area of scientific discovery. Here too there is a kind of "fetishism of invention," a taking of the product for the process, a neglect of the social and psychological elements that constitute discovery for the particuliar concrete product it creates. In this chapter we want to look at some of these social and psychological factors that make up the process of discovery so that we may understand the rate at which it can create products and the kind of products that are possible at any given time. Scientific discovery is not the mysterious outcome of unexplainable individual genius. It is rather the result of a partly specifiable social process in which the individual and society each has its important part to play.[1]

It is not strange, of course, that men should be victims of this "fetishism of inventions." While they have always known something in a common sense way of how to go about making a discovery, men have been far less interested in finding out *how* they were inventing than in finding *the particular thing* they wanted to invent. Their other purposes, their so-called "practical" purposes, have up to now usually been more important to them than the theoretical ones involved in generalizing their experience of discovery and invention. Not until perhaps the nineteenth century, as Whitehead has put it, did man make his greatest discovery, when he "invented the method of invention." We shall see that this statement is not wholly correct, for there is still a great deal we do not know about how discovery occurs; but only recently, in any case, have some men come to be as much interested in the social process of discovery as in its products.

There have also been reasons internal to science for the greater emphasis we have put upon its physical products than upon the social and psychological processes of discovery. There is, for instance, a strict convention in science that discoveries should be presented in their finished and rationalized form, with their logical structure and supporting evidence standing forth as clear and bare as possible. All else is considered distracting from the main purpose of science, which is the demonstration of the theoretical validity of a discovery. But this convention leaves out a great deal of what is of the greatest importance in science. It leaves out all the errors and all the fertile imagination of the scientific discoverer. "The raw materials out of which science is made," says the physiologist, René Dubos, biographer of Pasteur, "are not only the orbservations, experiments, and calculations of scientists, but also their urges, dreams, and follies."[2] It also leaves out the influences on each new discovery of what has gone before in science and what exists in the surrounding society. These are some of the things we shall need to look at if we want to understand the actual social process of scientific innovation.

In applied science, moreover, there are other reasons for leaving out a great deal of what occurs in the process of invention. In industrial laboratories, for example, the patent department reads all papers that are to be published, specifically in order to eliminate all unnecessary hints of the method used, this for the purpose of depriving commercial competitors of the valuable aid they could often thus obtain.[3] In both cases, of pure and applied science, further, most of the actual failures of research are simply not reported although the failures are sometimes as instructive as the successes for knowing what happens in scientific discovery. We shall shortly look at a case of failure which highlights the function of imagination in scientific discovery. Certainly the failures are very numerous in science, perhaps more numerous than the successes. All this is why there is no substitute for learning the method of science by actual first-hand experience in a laboratory, preferably as apprentice to someone who is skilled in what is often called the "art of discovery." The established substance of science can be learned pretty well through formal teaching. The method of science, or the "art" of discovery, like all "arts," is best learned informally in the master-apprentice fashion. "A master's daily labours will reveal," says M. Polanyi, "the way he chooses problems, selects a technique, reacts to new clues and to unforeseen difficulties, discusses other scientists' work, and keeps speculating all the time about a hundred possibilities which are never to materialize," and this "may transmit a reflection at least of his essential vision. This is why so often great scientists follow great masters as apprentices."[4]

What, then, are inventions and discoveries? First of all, we have to repeat that they have two aspects, that of process and that of product, and that these separate aspects must be distinguished, for otherwise a great many confusions of understanding result. And, second, we have to speak of their relation to ideas. In common parlance "invention" is used often to refer to a machine or some other physical thing; and, similarly, "discovery" is used to refer to a new set of ideas. This usage is not acceptable, since a new set of ideas underlies every discovery and every invention, and the particular concrete form in which the ideas are embodied is much less important than the ideas themselves. We can see this very easily when we are confronted by some new machine which we do not "understand," that is, for which we do not have the new ideas necessary for working it. This is also why tools and machines mean different things in different societies, and why, further, trained men have to accompany new machines to teach the ideas that give the machines their meaning and use.

We consider scientific "invention" and scientific "discovery" to be analytically the same, therefore, and we shall use them interchangeably. They may, then, be defined as the results of those imaginative combinations which men make of previously existing scientific elements in their cultural heritage and which have *emergent novelty* as combinations. This definition stresses, by its use of the term "imaginative," the role of ideas in invention, and also it indicates the similarity of scientific innovation to discovery in other cultural areas. Invention occurs among all the different types of ideas that compose the cultural heritage—ideas about nature, about social behavior, about aesthetic and artistic behavior, and about moral standards. Invention and discovery not only are not limited to science, but as they occur in science they are processes with a great many similarities, as well as differences, to innovation in the other fields of culture. The sociologist of science, for instance, has a great deal to learn about the processes of the human imagination from such a study as John Livingston Lowes' investigation of the poetry of Coleridge in *The Road to Xanadu.* Metaphor is not alien to the scientific imagination, although of course it has different functions from those it has in poetry. "In the last analysis," says Levy, the English mathematician, "there is little difference between the *individual* effort of the artist and the scientist in the direct handling of his problem. He who is devoid of imagination can be neither scientist nor artist."[5]

We need to make clear something else that is very important about scientific discovery. We are accustomed to think that only very grand sets of ideas and only very powerful machines, especially ones with far-reaching practical consequences, may be called discoveries and inventions. The overwhelmingly largest number of scientific innovations, however, are imaginative combinations which achieve only very small advances in novelty. Discovery is a process that is never absent from society; its innumerable manifestations usually make all but imperceptible contributions to the change and development of man's cultural heritage. Small scientific discoveries come about in essentially the same way as large ones, and in a sense they are no less important, for they are one class of the elements that must go into the large discoveries. Large and small innovations are necessarily intertwined.

Let us look for a moment at this matter of small inventions. Especially in a society like American society, which we have seen has so strong an approval of innovations of all kinds and has so many facilities for producing them, the number of inventions that occur, when the small ones are taken with the large, is immense. Of course many small inventions are not even patented, although our industrial technique finds them invaluable. "The Dennison Manufacturing Company, for example, employing about 3,000 men, received from its employees in the one year 1920 a total of 3,701 suggestions. Fifteen percent of all the suggestions received were adopted by the company."[6] This system of "suggestions" for innovations has spread widely in industry in the last thirty years. It was at its peak during the recent World War when there was an urgent necessity to tap every resource the country had of efficiency and inventiveness for the benefit of its industrial productivity. Other kinds of small inventions are even greater in number than those from this relatively casual source. For instance, there are already some two-and-a-half million patented inventions in the United States Patent Office alone, and most of them are very small novelties indeed. Take patents on the toothbrush. There are nearly one thousand patents on the toothbrush in the United States Patent Office. "Most of these 'inventions,'" says one student of the patent problem, "are modifications of the size and shape of the handle, or the number, size, and arrangement of bristles. In general, a relatively small proportion of all patents registered has greater technological importance than these patents on the toothbrush."[7]

The same thing must be true of the enormous number of small discoveries that are reported in scientific and technical journals every year. "As long ago as 1933," we are told, "S. C. Bradford, of the Science Museum Library in London, estimated that 750,000 scientific and technical papers were published annually. More recent estimates indicate that the rate has doubled since then."[8] This is the nature of an immensely prolific science. For the period 1917 to 1926, for example, the Second Decennial Index of *Chemical Abstracts* requires 6,600 pages of fine print.[9]

So small are most inventions and discoveries, which is to say, so small or obscure is the element of emergent novelty they contain, that it is often difficult and sometimes all but impossible to define for certain practical purposes what may fairly be called "an invention." The United States Patent Office, for example, suffers acutely with this difficulty. "We do not know," says one summary statement, "what 'invention' means."[10] The courts and the Patent Office have tried to use many different tests and definitions. "Rules of thumb have been applied and, almost as frequently, ignored, such as the rules that mere addition, or subtraction of elements, mere aggregation as distinguished from mere combination, changes in form, reversal of parts, substitution of elements, etc., do not constitute invention." More affirmative tests have also been rejected, tests like satisfaction to society, commercial success, and amount of research necessary. "In even more generalized terms, so indefinite as to afford little help as a workable test, invention has been described as 'something more than mere application of mechanical skill,' 'a flash of genius,' 'that impalpable something,' etc. When one is all through, there is little to do but throw up one's hands in despair and say that invention, like the Constitution, is what the judges say it is."[11]

The judges, of course, speak in some fashion for society, and this is another important thing we need to know about invention and discovery. Another essential component of our definition, we see, is that the emergent novelty of an invention or discovery must be socially recognized and socially rewarded. When they are held only by an individual and not by some social group, novelties are merely private fantasies which require communication and social acceptance to become inventions. In any society there are, of course, different groups to whom a novelty may be useful and acceptable. The number of groups is especially large in so highly differentiated a society as the United States, and that is one reason why there are so many discoveries and inventions. But some "inventions" do not seem to be useful to anyone. The infant mortality rate among even patented inventions is extremely high; many patents simply involve expense to their holder and are never used otherwise. A surprisingly large number of the inventions which are acceptable to the United States Patent Office are not also acceptable to any manufacturing company or to the consuming public. We should note, however, that the amount of imagination, or of the "inventive faculty" as it is often loosely named, may be as great in an innovation which is not socially useful as in one which is.

We have already referred to another social influence on invention, the influence of the existing scientific and technological heritage. The degree of novelty in an invention often appears to us greater than it actually is because the component antecedents from the culture heritage are less obvious than the form of the new combination. Especially to the outsider, the non-specialist, inventions tend to appear only as full-blown creations, with all their slow growth and evolution obscured by their present usefulness and success. But inventions and discoveries are in their very essence accumulations of previously existing elements, accumulations in which a degree of novelty may be present but may also be very small when the past is considered. "When one considers carefully the genesis of any discovery," says George Sarton, historian of science, "one finds that it was gradually prepared by a number of smaller ones, and the deeper one's investigation, the more intermediary stages are found."[12] The cumulative nature of discovery in science has been recognized for a long time. There is, for example, a saying ascribed to Bernard of Chartres, a scholar of the twelfth century: "In comparison with the ancients we are like dwarfs sitting on the shoulders of giants." A similar saying about his own work is often attributed to Newton.

Any novelty thus, necessarily is a "composite collective product," as Lewis Mumford has said of invention.[13] A scientific book, for instance, is composed at least of all the other books and articles to which it makes references, although these are only an extremely rough measure of its important components, as any scientist knows. So is it also with a machine. J. A. Hobson has pointed out that "the present spinning machinery which we now use is supposed to be a compound of about eight hundred inventions. The present carding machinery is a compound of about sixty patents."[14] The automobile is a whole series of inventions, similarly, the product of many thousands of patented novelties. Fortunately none of the patents is exclusive because of the cross-licensing agreements in the automobile industry. One of the greatest composite inventions of all, which we usually think of as a single invention or discovery, is the ship.[15]

Now it is sometimes asserted by those who recognize the importance of cultural antecedents in invention that the number of inventions that occur in a society will be greater the larger is the cultural heritage. But the existing culture base is only one of the social elements involved in the rate of invention, and moreover this assertion implies a certain social automatism about discovery which is not an accurate description of its nature. Common notions to the contrary notwithstanding, even Newton and Einstein were dependent on their scientific antecedents, yet this does not explain Newton and Einstein.[16] The rate of scientific advance also depends on the number of creatively imaginative individuals in a society. Elements in the cultural heritage do not spontaneously combine themselves into novel products. The cultural heritage only makes invention possible, not inevitable. In a little while, therefore, we shall look more closely at the role of the individual and his imagination in scientific discovery.

This sociological view of the importance of cultural antecedents in the process of scientific discovery is a valuable change from a conception of the nature of invention which formerly was much more prevalent than it now is. This older view, which lingers on in loose writing about science, may be called "the heroic theory" of invention, and it stressed the particular and peculiar genius of the inventor as against the contribution society itself made to his discovery.[17] In the eighteenth century, for example, it is reported, the Marquis de L'Hopital asked in full seriousness whether the great Newton ate and slept as other mortal beings did. The heroic theory was well suited to the simplicities of adulatory biographies and popular mythology, and especially to the enthusiasm of nationalistic patriotism. For example, whereas in fact several men "invented" the steamboat all at about the same time, the American may learn from his history textbook that Fulton invented it, the Englishman reads about his compatriot, Symington, and the Frenchman is told in school that Jouffroy was the real inventor. The heroic theory is, however, not nearly so commonly found now as it used to be, although the Russians have been reviving it lately in the interests of their greater national glory. For example, their propagandists, if not their scientists, are now claiming Russian priority in the invention of the radio, the airplane, the steam engine, and penicillin.[18]

We have just said that several men "invented" the steamboat all at about the same time. The steamboat is not the only case of this kind, that is, of what is called "independent multiple invention." Indeed, practically all students of the sociology of invention in recent years have noted this phenomenon, in which two or more men make the same discovery at approximately the same time without being aware that it is being made elsewhere or has already been made, and it has become one of the chief kinds of evidence for the sociological theory of discovery. It is a pattern that has already repeated itself a great many times in the history of science and technology and that continues to do so for reasons which we shall consider shortly. First let us see how extensive its occurrence is.

The sociologist, William F. Ogburn, was perhaps the first to draw up a list of cases of independent multiple invention.[19] Drawing upon the histories of astronomy, mathematics, chemistry, physics, electricity, physiology, biology, psychology, and practical mechanical inventions during the last few centuries, Ogburn found one hundred and forty-eight examples which involved two or more independent discoverers of the same thing. This was probably not a complete count even when it was made, some thirty years ago, and others have occurred since then. The following fourteen items are a few of those on Ogburn's list and show how wide is the range of discoveries which he includes:

Discovery of the planet Neptune	By Adams (1845), and Leverrier (1845)
Logarithms	By Burgi (1620), and Napier Briggs (1614)
Calculus	By Newton (1671), and Leibniz (1676)
Discovery of Oxygen	By Scheele (1774), and Priestley (1774)
Molecular theory	By Ampere (1814), and Avogadro (1811)
Photography	By Daguerre-Niepe (1839), and Talbot (1839)

Kinetic theory of gases	By Clausius (1850), and Rankine (1850)
Mechanical equivalent of heat	By Mayer (1842), Carnot (1830), Seguin (1839), and Joule (1840)
Telegraph	By Henry (1831), Morse (1837), Cooke-Wheatstone (1837), and Steinheil (1837)
Electric motor	By Dal Negro (1830), Henry (1831), Bourbonze and McGawley (1835)
Relation of micro-organisms to fermentation and putrefaction	By Latour (1837), and Schwann (1837)
Laws of heredity	By Mendel (1865), DeVries (1900), Correns (1900), and Tschermak (1900)
Balloon	By Montgolfier (1783), and Rittenhouse-Hopkins (1783)
Flying machine	By Wright (1895-1901), Langley (1893-1897), and others.
Reaper	By Hussey (1833), and McCormick (1834)

Just to mention another field, independent multiple invention is very frequent also in the history of medicine.[20] And one of the most recent cases of what we are speaking of is that of the use of a radio pulse technique to detect aircraft and ships. This technique, called radar in America and Britain, "seems to have occurred almost simultaneously to scientists in America, England, France, and Germany," says the historian of America's scientific effort in the recent war.[21] And finally, we may consult the patent record. Every year, so the records of the United States Patent Office show, thousands of inventions are re-invented.[22] Sometimes there is a lapse of time between the invention and its re-invention, but there are a great many instances in which inventions are made practically simultaneously by men living in different parts of the country and unknown to one another. Hence the frequent difficulty of determining who is *the* inventor of something. Hence also the frequent recourse to the patent proceeding known as an "interference," which is a legal hearing designed to name the prior inventor. For the period of the 1920's alone it is estimated that about four per cent of all patent applications represented independent multiple inventions.[23] One recent discussion of the problem says that "simultaneous invention is so regular as to be almost commonplace."[24]

It would be wrong to infer too much, as some have done, from the very frequent occurrence of independent multiple inventions. This phenomenon does not justify an extreme sociological determinism which sees scientific discoveries as products automatically thrown off by the impersonal movement of the historical process. It does demonstrate, however, that the body of scientific knowledge and technique is, at any given time, relatively structured, so that changes in it do not occur at random. Advance into novelty is in considerable degree selective because of the existing structure of scientific theory and knowledge. Of course this structure is not wholly autonomous; it is influenced, as we have seen in Chapter Two and elsewhere, by other parts of society: by values, and by religious, economic, and political factors.[25] In our next chapter we shall have still more to say about the social influences on discovery and invention. Yet there is also the inherent relative autonomy of the body of scientific theory; and from this as well as from social influence there emerge multiple discoveries by men whose activity is guided in part by the existing scientific heritage and in part by their creative imaginations.

In this connection we need to take warning against catchword descriptions of the social influences on scientific discovery. Formulas will not serve us here, formulas like "an invention must fit the times," or, "the times must be ripe for an invention," or "social need produces invention." Such statements are indeterminate; they beg the very questions we want answered. For instance, we know that "social need" does not always produce an invention, for many "social needs" have existed and still exist without calling forth adjustive inventions. The North American Indians "needed" coal and automobiles as much as modern Americans do, but they did not have the requisite scientific base, let alone all else that was needed, to produce these discoveries. Today we "need" a cure for cancer and many other things, but "needs" alone will not get us what we want.

The role of the individual in scientific research, however much his function and his particular problem may have been conditioned by society, is still an active one. Now this activity is often concealed by that impersonal and perfectionist way in which scientific discoveries are commonly reported. But reflective scientists have always taken note of the active part they play in making their discoveries.[26] Nature does not easily yield up finished conceptual schemes for the understanding of the relationships that hold among its parts. The scientist, always making full use of the already existing conceptual schemes, in addition always actively puts questions to those parts of Nature about which he hopes to discover some new relationships. The next questions in science are never obvious, moreover, never equally apparent to all men. The successful scientific discoverer always uses what René Dubos calls "anticipatory ideas," that is, questions which he himself actively constructs and then submits to test by experiment. Only in the course of the experiment itself does the scientist follow the ideal of passively observing results. If these do not conform to his expectations, based on his "anticipatory idea" or hypothesis, then he makes another active anticipation, he forms another hypothesis for experimental test. "It often happens," says the great Claude Bernard, "that an unsuccessful experiment may produce an excellent observation. There are, therefore, no unsuccessful experiments."[27]

In all of this activity, the individual researcher must exercise as much creative imagination as he can to see newly significant connections between existing elements of theory and knowledge. Only by imagination in forming hypotheses does emergent novelty, or scientific discovery, occur. The place of imagination in scientific discovery is particularly apparent in "the flash of insight" which occurs to all creative minds. One of the best known cases of this kind is the "sudden flash" of intuition by which, as Darwin reports it in a letter to a colleague, his theory first occurred to him in 1844. A great many other scientists as well have recounted experiences in which they had sudden "hunches," "flashes of insight," and "intuition" into possible relations that they had never seen before.[28] The great chemist, Kekule, originator of structural formulae in organic chemistry and discoverer of the benzene ring, was speaking of these things when he said, "Let us learn to dream, gentlemen, then perhaps we shall find the truth." He meant "dream" literally as well as figuratively, for many scientists have had some of their greatest insights while they slept. These sudden perceptions of new "wholes" do not, however, come entirely out of the blue; they usually occur only after a scientist has been long preoccupied with some problem.

Another place where we can see the function of active imagination in the process of discovery very clearly is what has come to be called "the serendipity pattern." The physiologist, W. B. Cannon, has defined "serendipity" as "the happy faculty, or luck, of finding unforeseen evidence of one's ideas, or, with surprise, coming upon new objects or relations which were not being sought."[29] Cannon tells us in his scientific autobiography that "during nearly five decades of scientific experimenting instances of serendipity have several times been my good fortune."[30] For example, his important discovery of sympathin was due to good luck of this kind. The occurrence of the serendipity pattern, like that of independent multiple invention, has recently been remarked by a great many practicing scientists and by those who study their activities. There has happened, in this respect, so to speak, another case of independent multiple discovery, this time about the nature of the process of scientific discovery.[31] The following are a few representative cases from among the many which have been noted: —Galvani's discovery of the electric current; Claude Bernard's discovery of animal glyco-genesis; Roentgen's discovery of X-Rays; Charles Richet's discovery of allergies; Alexander Fleming's discovery of the antibiotic effect of penicillin; Pasteur's work on immunization and crystalline structure; William Beaumont's work on the digestive processes; Dam's discovery of Vitamin K; Goodyear's vulcanization of rubber; Nobel's invention of dynamite; and Perkin's first synthesis of an aniline dye. Indeed, Ernst Mach had listed a great many cases as early as 1896. "Under this head," he said, "belong the first disclosures of electrical and magnetic phenomena, Grimaldi's observation of interference, Arago's discovery of the increased check suffered by a magnetic needle vibrating in a handbox, Foucault's observation of the stability of the plane of vibration of a rod accidentally struck while rotating in a turning-lathe, Mayer's observation of the increased redness of venous blood in the tropics, Kirchhoff's observation of the augmentation of the D-line in the solar spectrum by the interposition of a sodium lamp, Schönbein's discovery of ozone from the phosphoric smell emitted on the disruption of air by electric sparks, and a host of others. All these facts, of which unquestionably many were *seen* numbers of times before they were *noticed,* are examples of the inauguration of momentous discoveries by accidental circumstances and place the importance of strained attention in a brilliant light."[32]

Also, we may say, the difference between *seeing* and *noticing,* as Mach puts it, highlights the importance of individual creative imagination. These "unexpected" occurrences in the pattern of serendipity have been passively seen by other scientists; they are actively noticed only by the discoverer. They are actively noticed, that is, by the scientist who has carefully studied his problem over a long time and is thereby ready, if he can create some anticipatory ideas, to take advantage of an "unexpected" occurrence. Pasteur long ago expressed this essential pre-condition of serendipity in a classic statement. Chance, he said, favors only the prepared mind. Of course, even after the active construction of an hypothesis, as we have said, discovery is not completed. There remains the experimental test which validates or invalidates the expected relationship. We say "or invalidates" because surely there have been innumerable cases of what we may call "negative serendipity," or the chance perception of *apparent* connections among things which have not stood up when put to the experimental test.

It often happens in science that after a discovery has been made "by chance" other scientists remember that they have in the past passively seen what someone has now actively noticed. The industrial scientist, F. R. Bichowsky, has recorded a case of this kind, in which his imagination failed to take advantage of what chance presented to him.[33] "Back in 1912-13," he reports, "Sir William Ramsey, the discoverer of argon, neon, krypton, and xenon, lectured before the Lowell Institute of Boston. I helped in the preparation of the experimental demonstrations for these lectures." Ramsey, who was a brilliant lecturer, took three or four lectures to tell of his discovery of these inert gases, illustrating with experiments. He told how he named the gases neon, meaning new, argon, meaning lazy, krypton, meaning hidden, and xenon, meaning strange. He told how he had tried to combine these gases with other substances, but he said that they remained useless scientific curiosities because they did not so combine. In his last lecture, he said, "Some of you will ask how we can be sure that these gases are really pure substances, not just mixtures. I will show you. All pure substances are characterized by the fact that they give out, under an electrical discharge, their own special kind of light." Then he passed an electrical charge through a series of tubes and each one lighted up with the pale glow of a distinctly different color. "Under different conditions of discharge," he then said, "these colors can be intensified." Thereupon he switched on a condenser in the line and the tube containing neon flashed up a brilliant orange red light. "It was very striking," says Bichowsky. "We all applauded and went home. Not one of the five hundred or so who heard the lecture realized that we had seen the first Neon sign. It was only some years later that Claude, seeing exactly the same experiment, realized its commercial importance."[34]

The serendipity pattern only underlines a general fact that is of primary significance to the understanding of science. When one views the course of scientific discovery through the careers of its individual researchers, and not just as an impersonal series of events, its progress does not appear inevitably determined to go in some very particular direction, as we might expect on the view that the social process of invention is automatic. Looking back on them, we feel certain that Pasteur's discoveries, for example, have a definite logic in their sequence. But this logic, says Dubos, "was not inescapable. His career might have followed many courses, each one of them as logical, and as compatible with the science of his time." And Dubos has himself shown quite specifically just what these other courses might have been.[35] The social influences on science set limited alternatives for the individual, but still they are alternatives, not predetermined tracks.

With all this understood, then, we can perhaps see more clearly the meaning of that epigram of Whitehead's to which we referred earlier, that "the greatest invention of the nineteenth century was the invention of the method of invention." We have not so much invented the process of discovery in the sense that we understand it fully or can control it completely, as we have created some of the conditions in which invention can occur more frequently. We have a great body of professional university and Government scientists and industrial research workers, all devoting their entire efforts to discovery; they are imaginative and are equipped with increasingly powerful and extensive conceptual schemes, instead of with relatively empirical, trial-and-error methods; and we have a society which actively favors the development and use of scientific innovations. Under these conditions, the social process of invention and discovery flourishes as it never has before in the history of human society.

X

The Social Control of Science

IT IS NOW a commonplace to say that science has vast and far-reaching social consequences. It is a commonplace, this fundamental fact, because we are every last one of us aware, at least since the explosion of the atom bomb, that it is so. There is no longer any place to hide from this fact. But this does not mean that it is an essentially new fact. Rational knowledge and science, and their applications more directly, have always had important social consequences, in all societies. They have always had their effects upon other parts of the society as much as they themselves were influenced by these other parts. This ever-present reciprocal relationship between science and the rest of the society, we saw in Chapter Two, was a basic theme for our understanding. In the last three hundred years, however, as highly developed science based on general and systematic conceptual schemes has produced a continuous stream of discoveries and inventions, the rate and force of the social effects of science have been multiplied in geometric proportion. As a result, an old fact has taken on new significance, and even seems to be a wholly new fact.

Now in the past, many of the social consequences of science have been indirect rather than direct; they have operated only through other social factors than science itself. For example, many of the effects of science during the last few hundred years have appeared in the form of new social arrangements in industry and new technology. As a result of the indirect workings of science, many men were not aware of their ultimate source at all, or they could ignore it because social change did not work to their disadvantage. Science has had an unequal effect upon different groups in modern society, for instance, upon industrial workers as compared with middle class *rentiers*. The former were usually too taken up with the immediate impact upon their lives to look farther for its causes; the latter did not feel the need to become self-conscious about what so steadily benefited them. Of course a few men, social theorists and social reformers, did see some of the particular effects of science on society and the general significance of its new force as well. But their predictions of impending scientific millennia or their warnings about dire social consequences went largely unheeded by the mass of men, perhaps even more so by scientists than by others. The days of our blindness are now past, though. No one can any longer ignore the meaning science has for the present and future shapes of American and world society.

All this has raised the problem of the social control of science in a new way. That is one of the reasons why men seek to understand this wonderful modern beast which harms them only a little less, they sometimes think, than it helps them. While some see in science the solution for all the troubles that beset us, others see in it the source of most evil. On the one hand there is talk of "the frustration of science" and of "the need for planning science," and on the other hand men ask for "a moratorium on invention and discovery." Science has become for many of us a "social problem," like poverty and juvenile delinquency, and men want to "do something about it."

What shall we do? We shall do nothing well, of course, unless we really do understand science and the nature of its social consequences. In previous chapters we have been trying to get some preliminary understanding of this kind; and now, in this chapter, we want to relate some of the things we have already said to this problem of the social control of science. The matter as a whole actually involves several different questions, and we shall look at some of them here to see what light we can throw on this aspect of science, that is, on science as a social problem about which men have strong moral feelings and for which they recommend radical action. Are the social consequences of science inevitable and uncontrollable? How are the effects of science already controlled by other parts of society? What are the "resistances," as some people call them, to science in modern society? Since science does have differential effects, as we have said, on which groups does it act beneficially, on which harmfully? Can we say that science always has only harmful or only beneficial consequences for a given social group? Can we predict inventions and discoveries, and thereby control the effects of what we can foresee? Are we likely to stifle science, or frustrate it if we limit its consequences? Science is only one among the several social values we hold dear. What is the effect of science upon our other social values? Need there be perennial conflict between science and some of these other values, say, the humanitarian ones? What "social responsibilities" do the scientists themselves have to deal with the social problems their activities create? Can science be "planned" in such a way as to have it do only what we want? Can science be "planned" at all?

These are a great many questions, and there is no final answer to any of them, no absolute solution for the social problems they describe. We shall say what little we can about them in three sections in this chapter: A. The social consequences of science; B. The social responsibilities of science; and, C. Can science be planned?

A. The Social Consequences of Science

IN HUMAN SOCIETY, social change is only a matter of degree. No matter how relatively "static" some societies may appear in comparison with others, all of them undergo continual change. Some of this change is the result of things external to a society, things like other societies or like the physical environment; and some of it is a consequence of internal changes in things like the cultural values or the knowledge or the social arrangements of the society itself. In modern industrial society, and most all perhaps in American society, not only is social change continual but its pace is extremely rapid. Some of this change still arises in the external situation, in such matters as the emergence of a great new power among the nations or the discovery of a valuable natural resource. But a large part of the change in modern society is now inherent in its own internal nature, in the essential conditions of the functioning of industrial society itself. This is the basic truth we have in mind when we say that we live in a "dynamic society."

One of the chief internal sources of social change in modern society is science and its extensive applications in industrial and social technology. By our approval of science, by the way in which we provide such large opportunities for those who want to work at science, there has been introduced into the very heart of our society a fundamentally and continually dynamic element, an element which must remain the fount of unending social consequences, for both "good" and "bad." Note, for example, the roots of this fact in the attitudes of the scientists themselves, who feel this way only a little more strongly perhaps than the rest of us. The following question was asked of a representative group of scientists in a Fortune Magazine poll: "Check whether you believe a scientist should (1) withhold a discovery from the world when convinced it would be productive of more evil than good, or (2) never withhold a discovery, leaving it to the moral sense of mankind to decide its ultimate use." 78% of the university scientists, 81% of the Government scientists, and 78% of the industrial scientists answer that they would "never withhold" a discovery, whatever the consequences.[1]

We face a new condition in human society. The simple truth is that we must learn to live with social change because we value very highly that which cannot do otherwise than cause change. Of course we may, either deliberately or unwittingly, decide we do not like so much change and the continual social consequences of science. If we do so decide, then we must cut off change near its roots; we must restrict science a great deal more even than we now do. We cannot, however, have both science and complete social stability. The price, moreover, for greatly restricting science would be to have a greatly different society. For not only do our values approve of science, but all our social arrangements, as we have indicated in Chapter Three and elsewhere, are integrated with its successful functioning. An industrial society can no longer maintain and increase its prosperity or its power without science and its applications. "Our whole economy," a distinguished economist points out, is "geared to a rapid rate of change, and a drop in expansion or replacement means depression."[2] Nowadays we do not so much fear depression as war, and a drop in our power and our science could mean war as it formerly meant economic depression. We cannot restrict science and its social consequences in any absolute way lest we vitally reduce the viability of our society in a dynamic world in which science is the mainspring of social stability as well as change.

The process of continual change we have been speaking of is part of what Max Weber was referring to in his discussion of "the process of rationalization" in the modern world.[3] Weber had in mind not merely the changes deriving directly from the natural sciences but also those from the social sciences and indeed those from the whole ramified structure and application of rational thought and activity in our society. We have already said that science in our society is dependent upon the value placed upon critical rationality throughout the society. The rationality of science is only the sharpest instrument of this value of ours and only the most fruitful source of social consequences. But rationality, wherever manifested, has the same effect of producing changes and of undermining established social routines. Social instability is in part, then, the price we pay for our institutionalization of rationality.

Social instability and its consequences are not something to be treated lightly. It is not strange that they should be the cause of that ambivalence toward rationality in general and science in particular which seems to be widespread, though usually latent, in our society. The products of "the process of rationalization" do not have single effects but rather diverse ones for different groups in the society. All of us are pleased with some manifestations of rationality and not others, pleased with some of the products of science and not others. All of us are rendered sometimes more insecure, sometimes less, by the changes these things bring about. The standing routines and the vested interests of every member of society are many times attacked and overthrown by "the process of rationalization." No doubt the feelings of hostility and uneasiness which result are usually counterbalanced and even outweighed by the favorable consequences of science that all of us also experience. But there remains in each of us a residue of ambivalence. And it is this residue, together with the still stronger hostility which some men have against "the process of rationalization," which have been mobilized by agitators and dictators in the modern world. As they and their followers see it, capitalism and bolshevism and science alike are evil consequences of "the process of rationalization." No wonder Hitler seemed to make sense to a great many Germans when he lumped them all together. He even seemed to make sense when he made the Jews the symbol of what lay behind all three and the scapegoat for its attendant consequences.

Perhaps the analysis of the nature of science and its inevitable social consequences which we have been developing here will permit us to see the "technological theory" of social change in a new light. This theory, which has had a great vogue among some social scientists, holds that it is change in technology which always produces change in the rest of the society. The theory was most epigrammatically stated by Veblen when he said that "invention is the mother of necessity."[4] The technological theory of change, we can now see, does not search far enough. If one wishes to trace the source of social change no farther back than the technological innovations that are the product of science, then the theory does hold when it says these innovations are an important source of change. But behind the technological and social innovations lies the primary source, science itself, dynamic by its very nature and continually producing not only new conceptual schemes but also the possibility of new applications of those schemes in the form of technological inventions.

And of course the theory of technological change does not hold insofar as it may be taken to assert that technology is the *only* source of social change. By the same token, it would be incorrect to say that science is the only source of social change. Science and technology, we have seen many times now, are in interaction with the other important parts of society, and therefore they are sometimes dependent variables as well as sometimes independent variables. This does not mean that we must always trace a series of social influences through all its ramifications. For some purposes it may be enough to stop at some intermediate point. For some purposes it may be enough to stop with technology as the source of social change. But for other purposes it is not; for example, if we wish to maintain the flow of technology itself. Wherever one stops, it is well to know what one is ignoring; for unknown variables have a way of exerting their effects in uncontrollable fashion.

The interdependence of science and technology with other social elements can readily be seen in the existence of what those who explicitly or implicitly accept the technological theory of change have called "resistance" to discovery and invention.[5] When science and technology produce innovations which might cause changes in other parts of society, these scientific discoveries are not always nor automatically put to use. From the point of view of the technological theory, this failure to employ innovations seems like "resistance" to a powerful social agent which will inevitably have its way. But when we consider the partial dependence of science and technology upon society, and when we consider how the component factors of society are in interaction, we see that these "resistances" can also be taken as indications of the relative autonomy of other social elements than science and technology. As we might expect, therefore, "resistance" to innovation occurs in all societies, not just our own, for there is always some group for which a specific discovery is at best of no use and at worst an obvious threat of harm.

Professor B. J. Stern, who has been the leading student of this problem of "resistances" to invention, has given a long and typical list of some instances. "These resistances," he says, "have not been exceptional, but have generally characterized the response to innovation. The railroad, automobile, street car, steamboat, iron ship, screw propellor, submarine, airplane, typewriter, telegraph, telephone, cable, steam engine, Diesel engine, gas for lighting, incandescent lamp, alternating current, important processes in the manufacture of iron and steel and of textiles, the sewing machine, the iron plow, mechanical planting and threshing machines, tractors, the cotton gin and mechanical cotton picker—these are but a few of the important innovations upon which modern living rests that have met opposition of varying degrees."[6] The very length of this list should perhaps have warned us that a great many things were being lumped together which could more profitably be taken separately. We need to translate the term "resistances" into the phrase "interaction with other social factors" so that we can search out some of these other factors and see what influence they have on scientific discovery and invention. Toward this end let us look in a little detail at just four such factors: the needs of a going social system, certain moral and humanitarian values, the economic interests of established business enterprises, and the social and economic interests of industrial workmen in their jobs.

One of the fundamental characteristics of modern social organization is the pattern of standardization, a pattern which expresses itself most completely in machine technology but which also exists in social technology. Many of the benefits of the large-scale industrial mode of production in our society inhere in this pattern, and therefore some innovations which make greater standardization possible are highly welcome.[7] But once standardization has been established in some particular way, other innovations are much less welcome. That is, standardization has a double significance, at once encouraging scientific invention and acting as a brake on it. For example, take the case of our railroad system. There is probably somewhere around $100. billion invested in the American railroad system. But more than this, more than being only a capital investment, American railroads are themselves a going social organization, the functioning of which is bound up in an already standardized rail-gauge, standardized signals, standardized tunnel clearances, standardized rolling stock, and other kinds of uniform equipment. And they are indispensable to American society as a going social system. Now given this situation, one-hundred-and-fifty-mile-an-hour mono-railways which could cross an abyss on a steel cable might offer advantages in speed and efficiency over the existing railroads, but also they would greatly disrupt a going system. The resultant disadvantages are sufficient, especially when taken together with the probable losses to invested capital, to stifle the introduction of this particular technological invention. Now in all societies, and most of all in modern industrial societies—capitalistic, communistic, or otherwise—situations of this kind will occur many times on a large and a small scale. As a result, in all societies, even Communistic societies, there will be "resistances," necessarily, to changes in standardized equipment which is geared into the vital functioning of the going social system.

The history of science includes a great many cases of opposition to its innovations by vested interests in certain moral or humanitarian values. For example, people with certain moral convictions about the sanctity of the human body were opposed to the dissection of corpses and the performance of autopsies until well into the nineteenth century, and their opposition considerably impeded the advance of biological and medical science. The Commonwealth of Massachusetts was the first English-speaking community to legalize the dissection of cadavers. This it did in 1831. Not until the next year, 1832, did England legalize this practice. Before that time, medical men and scientists illegally procured cadavers from the so-called "sack-'em-up" men. Most American states did not legalize dissection until after the Civil War.[8] The moral and emotional congeners of those opposed to scientific dissection were, in the later part of the nineteenth century, the animal anti-vivisectionists. These latter, of course, are still with us in small numbers, and they are powerful enough in some states to require medical scientists to combat them actively from time to time lest they succeed in having anti-vivisection legislation enacted.[9] All this is a waste of time for the advance of biological science, but fortunately it is still only a nuisance and not a genuine threat. There are also more general kinds of moral opposition to science and its consequences. For instance, because his humanitarian values were disturbed by what he thought were the harmful social consequences of modern technology, the Bishop of Ripon advocated a "moratorium on invention" at the meetings of the British Association for the Advancement of Science in 1927. Such recommendations were heard again, this time in the American Congress, during the Depression of the 1930's, but they were never able to gather sufficient support to take actual shape as proposed legislation. Whatever their humanitarian or other moral origin, such policies were too obviously a counsel of final despair in an industrial society. Only in a utopia is an absolute prohibition on invention even thinkable. So, at least, did Samuel Butler construct his *Erewhon,* with a prohibition on any innovation that might disrupt social stability. Literary utopias, however, do not have to face up to the practical consequences of their absolute social choices.

It used to be alleged that the largest "resistance" to science in modern industrial society consisted of the suppression of invention by capitalistic financial or manufacturing interests, especially those seeking to develop or maintain a monopoly.[10] Now the evidence for this supposed suppression of invention was the unquestionable fact that a great many industrial firms held large numbers of patents which they did not put to use. Some of these patents may indeed have been held back against the general public welfare. But the evidence now seems pretty clear that very few of the unused patents were being withheld for merely narrow, selfish interests.[11] We have seen in the last chapter that only a few of all the patents that are taken out ever get used. It is the precautionary practice of firms that do research to take out patents on everything they develop, good or bad. But less than one per cent, for example, of the patents held by General Motors Corporation have proved useful. Or, to give another case, the great majority of Ingersoll-Rand Company's patents have never been used.[12] Many patents are highly obsolescent and are useful only until a better device can be made and patented. Some patents are unusable because co-ordinate technical development is lacking; they are therefore held in the hope that the other things they require will be forthcoming in research. We have already said that many patented devices and practices are not acceptable to their potential purchasing publics. And, perhaps most important of all, it is unwise to assume that there is something intrinsically economical about every discovery and invention. The benefit to society of some inventions is less than already existing devices that meet the same purpose. Or, if equal in benefit, the capital cost of a new device may make it less economical[13] The capital cost and obsolescence factors, says Lord Stamp, the English economist, "cannot be spirited away." No matter how much a society approves of scientific innovation, whether that society is capitalistic or socialistic, it has to calculate the economic and social costs of the new products science provides. "'Supression of patents' is not," says one student of the matter, "only a matter of patents. It is related more fundamentally to the pecuniary and social cost of change."[14]

The last social factor we want to consider as a "resistance" to science and its application is the fear that industrial workmen in modern society have always had that new machines would result in technological unemployment for them. Because of their relatively weak position in an industrial system, the workers have often borne the most directly harmful effects of technological innovations which have been beneficial to the capitalist groups and probably also to society taken as a whole. This opposition of the workers to technological and social innovations in industry has been called by Spengler "the mutiny of the Hands against their destiny."[15] Certainly this has been the destiny of industrial workmen for a long time now. The history of the last three or four centuries contains a continuous series of cases in which workmen, fearing technological unemployment soon or late, have opposed the introduction of new machines into their work situation.[16] David Ricardo, for one, did not think they were entirely wrong to be afraid. "The opinion entertained by the laboring class that the employment of machinery is frequently detrimental to their interests," he said, "is not founded on prejudice and error, but is conformable to the correct principles of political economy."[17]

This is not to say that "resistance" to new machines or new work routines is absolute among workmen. It varies as workers feel themselves more or less threatened by future possibilities. Of course, insofar as they are familiar with the history and institutional imperatives of the capitalist form of enterprise, imperatives which push toward continual technological innovation, workers will be subject to a universal, if latent, fear that job displacement is an ever-present threat. There is no reason to believe that workers do not know about such things. There is no reason to believe, for instance, that workers are entirely in the dark about the possibility of "automatic" factories which will be run by a handful of men. Managers may hear sooner and more clearly of these things, but the mass media, workers' unions, and informal story-telling among themselves serve the workers in a similar, if cruder fashion than management journals like Fortune Magazine.[18] Hence the various and subtle devices which industrial workmen use to protect their jobs against too much and too rapid innovation. All the techniques that make "restriction of output" possible, when seen in this light, have a useful function for the worker who has only his labor power to sell in an industrial society. Because of the crucial significance of the job in our society, labor displacement is virtual social displacement for the individuals affected.[19] Hence the urgency for workers of opposing the machine and other organizational innovations which are the immediate agencies of job displacement.[20]

Now the economic and social hurt the technologically displaced worker suffers is immediate and in the short run, so to speak. This hurt is not much alleviated by the general and impersonal observation, sometimes offered to him, that *in the long run* science and technology are good for the society, that in the long run they produce a greater volume of employment and a higher level of general well-being. Lord Stamp, in his Presidential address to the British Association for the Advancement of Science in 1936, when the problem of technological unemployment was much more pressing than it is now, said: "It does all this 'in the long run,' but man has to live in 'the short run,' and at any given moment there may be such an aggregation of unadjusted 'short runs' as to amount to a real social hardship." The industrial worker in modern society does not have his own capital to sustain him over the long run, as do industrial enterprises. His whole life is lived on the margin of his current earnings, and therefore he has to live in the short run. In recent times, various forms of "socialized capital" have helped sustain him. Government and trade union unemployment insurance benefits now help the worker over short run difficulties arising in technological or other kinds of joblessness. In such a situation, the modern industrial workman can afford to be more hospitable to innovation in his work situation. Otherwise he may turn his hostility against the machines which are the tangible source ot his grievances, as did the Luddite rioters in early nineteenth century England, or he may turn against their more remote source in science itself.

These, then, are a few of the social factors which have interacted with scientific innovation and have in that sense controlled its effects. Those who interpret this interaction as "resistance" imply that science is too much controlled or controlled by the wrong things. But other men in modern times have felt that science has not been controlled enough. The Bishop of Ripon's call for a "moratorium on invention" was probably only a strong statement of this feeling. There has also been another point of view about the control of science. This is the view that science could be controlled more than it now is, if only we could predict the emergence of important scientific discoveries and forecast their social consequences. In this way society could use science as it wished. Acting upon this view, several social scientists have very carefully tried to predict the course of science and its social consequences.[21]

A word about prediction in general before considering the specific predictions that have been made about scientific discovery. It is the assumption of our whole effort to understand the social aspects of science that predictions about social life are possible because the several parts of a society are related in determinate ways which we can know. Prediction is the statement of some of these determinate relationships with regard to some specific empirical social situation. We shall say more of this possibility of prediction when we discuss social science in our next chapter. Here we want merely to state our assumption so that the critique we shall offer of the existing attempts to make particular predictions about the course of science will not be taken as a critique of prediction in principle. It is desirable to show why these early efforts at prediction have been inadequate only in order to move on to more scientific predictions not only about the consequences of science but about all social life.

Let us consider first the problem of predicting scientific discoveries, in which what we said in the last chapter should be of some help to us. Insofar as particular discoveries and inventions are genuine novelties, there can of course be no absolutely certain prediction of whether or when these novelties will appear. This is only to repeat that the advance of science is not absolutely determined. Innovations are themselves, so to speak, cases of successful prediction by scientists and inventors of what can occur. If someone else than the inventor could predict the particular novelty, he would himself be the inventor. In *Technological Trends and National Policy,* for example, "there was a surprising failure to predict, or even to note the possibility of, some of the more radical inventions which were just around the corner. The section on air transport, for example, makes no mention of jet propulsion or of helicopters, although both were then under active experimentation and current results were good enough to require at least very careful consideration."[22] Nevertheless, because novelties are in considerable part a product of the scientific heritage, it is possible to predict *likely* occurrences deriving from the existing structure of scientific knowledge. Thus, for example, it was possible to predict in 1920 that it was likely that science would one day discover how to transmute one element into another and thereby release atomic energy. It was at that time that Sir Daniel Soddy said, "Whether it takes years or centuries, artificial transmutation and the rendering available of a supply of energy as much beyond that of fuel as the latter is beyond brute energy will eventually be effected."[23] Or, to come down to more recent times, it is now possible to predict that important discoveries are likely in the field of photosynthesis. In this connection the chemist, Farrington Daniels has said, "There is being accumulated rather rapidly now, along several different lines, a considerable stock of fundamental facts which should lead to a rapid unfolding of our understanding of photosynthesis."[24] Considering the whole matter of whether prediction of discoveries can become a science, Samuel Lilley, the English scientist, says: "The moral for forecasters is: Do not predict individual inventions in detail—that is usually a waste of time. Concentrate on two things; first, the extrapolation of present trends (for example, the further development of already existing petrol-driven land transport . . .); second, predictions of the form, 'it will become possible to fly.' "[25] Both Soddy and Daniels have complied with this advice when they limit their predictions with phrases like "whether it takes years or centuries" and "should lead to a rapid unfolding."

So far as the second problem of prediction is concerned, that of forecasting the social consequences of some particular discovery, two new difficulties appear. There is first the difficulty of detecting among the large number of emergent novelties which exist at any time those which are likely to be developed or to have important social consequences. This difficulty has been referred to, but only in passing by Professor Stern, himself one of the forecasters writing in *Technological Trends and National Policy.* "The annals of invention," he says, "are crowded with innovations, originally hailed as epoch-making, that have come to naught."[26] It is hard even to know what has already occurred. No single scientist can be expert enough in every field of science or indeed even in a whole "field" of science, say one so large as physics or chemistry, to know all the most recent important discoveries. The biologist cannot be expected to know what is happening at the frontiers of knowledge in chemistry, nor the physicist in similar areas of biology. The detection and reporting of important novelties in the different fields of science must at its best be carried on by a whole corps of alert specialists. Even such a group, however, would have its limitations. For scientists are like other men, though perhaps less so in this respect, in their tendency to enthusiasm about the significance and potential consequences of those discoveries they see most closely. In a world of specialists, where each expert loses perspective because of his "trained incapacity" to relate himself to other specialists, scientists are inevitably prone toward exaggeration of their own technical specialty.

There is, second, a difficulty in prediction even when an innovation has progressed to the point where it is obviously going to have far-reaching social effects. There is great difficulty, for example, in predicting, as Professor Ogburn has tried to do, the social effects of the airplane, an obviously important invention.[27] We can say with him, in a pretty indeterminate fashion, that the airplane *will* have important social consequences, but we cannot proceed very far toward determinate and usable statements about just what these consequences will be. Why should this be so? It is so because the development of the airplane and its use is related to a great many different social factors, and we cannot assume that these factors will remain constant, as we can, say, in the case of the different social factors which affect the trend of the birth rate in the United States. We can extrapolate the latter trend, because of the constancy of the relevant social conditions, even if we do not know all about them, but we cannot do the same thing for the airplane because of the great variability in the social factors that increase or impede its use.[28] In the case of the airplane the action and reaction of social conditions allows more scope for alternative courses of development; there is more room for unpredictable "resistances."

This is not to say, however, that the course of development is wholly indeterminate but only that it is less determinate, by far, than something like the population trend. Indeed, there are very few trends in social life which we can forecast with even the degree of relative determinacy which holds for population growth. Take another important discovery, that of atomic energy. "It must be confessed," say Newman and Miller in their book on atomic energy, "that the prognosticating powers of the best-informed nuclear scientists and the most perceptive social scientists with respect to future developments in the science of nuclear physics and the social, political, and economic effects of such developments are not impressive. The members of the Senate Special Committee had little but generalities (read: highly indeterminate statements) to help them in drafting legislation to meet a new technological era and possible revolutionary changes in social institutions. An analysis of the testimony offered reveals little more about the nature of impending changes than the conviction that they will come and that they will be important."[29] Obviously we are not yet capable of long-range forecasts about important discoveries which are going to affect the whole society. But something else important is left to us. We can still adopt the more modest goal of using what knowledge we have to make a *continual series of short-range forecasts,* and perhaps thereby we can even build up our knowledge so that we can eventually make predictions of increasingly extended range. This is not, let us note, a counsel of despair, but rather a recommendation that cautious optimism about the possibilities of useful prediction of the social consequences of discovery and invention will, at this stage of our knowledge, serve us best.

In this connection we must attend to a special characteristic of social prediction which will predispose us all the more to making short-range rather than long-range forecasts. It is a special condition of social life, that is, as against physical and biological phenomena, that predictions themselves become a part of the interacting set of social conditions which affect the development and consequences of scientific innovations. Thus the prediction that a certain discovery will have a given effect may actually stimulate the realization of that effect. Insofar as it does, it is what has been called a "self-fulfilling prophecy."[30] Or, contrariwise, prediction that a certain effect is possible may stimulate "resistances" which slow up its realization or even abort it altogether. This effect of social prediction has been called "suicidal prophecy." Not all predictions have equal potency in the social process, of course. Some may become major factors in altering the course of history. It has been said, for example, that the Marxian predictions were of this major order of significance, that they may have forestalled revolution, at least in Germany and England, where Marx predicted its outbreak. Most social predictions seem to be of a much lower order of effect in their interactions with other social conditions. Yet because of these two different consequences of predictions themselves, we do better to make continual short-range predictions in social life. Only thus can we constantly assess the new situations which exist because of the interaction not only of the previously existing factors but now, also, of the predictions themselves.

We have been speaking, up to this point, of matters having to do with further controls on science and its social consequences. Now we have to consider a group of scientists who have thought that science has been too much controlled in the modern world, especially in capitalist society. This is the group of so-called "scientific humanists" who have complained of "the frustration of science."[31] The group was mostly composed of a number of British scientists, many of them quite distinguished, but there were also some Americans who shared its attitudes.[32] The "scientific humanists," chiefly men of socialist philosophy, were anxious to do good to suffering humanity by all the devices and possibilities of science. They knew at first hand the actual and potential power of science to do good, and, knowing this, wished to maximize the realizations of that power. But in the pursuit of this ideal, which has been shared by many other scientists, they tended to absolutize science as a value in itself, or at least they ignored its interdependence with many other needs and values of a society.[33] The "scientific humanists" saw all restrictions of the full potential of science and invention as evils. Specifically, since they were socialists, they saw them as capitalist evils. They did not see that many "resistances" to science in our society come from other than capitalist economic interests, as we have just been suggesting. They did not see that there would inevitably be restrictions on science in all societies, since science can always be only one among several important social goals and must therefore share the available social resources of men and materials with these other goals. It cannot be assumed that there would be no restrictions on science even in a socialist society, although it is possible that there might be fewer than in capitalist society. But this possibility cannot be taken as a certainty, as some of the "scientific humanists" have learned in recent years from their former ideal, the Soviet Union. Certainly "liberal" capitalist society has, up to now, offered a remarkably favorable social environment for the advance of science.

All this apart, "scientific humanism" offers a view of science that we have not had very much of in the modern world, at least until quite recently, and that is the view of scientists asserting the *morality* of their activity in opposition to its "frustration." A great misconception of our society has been to think that science is wholly a-moral, and the scientists themselves used to take this stand as much as laymen did. But in "scientific humanism" we see what we stressed earlier, in Chapter Four, that science rests on a definite set of moral values and that these values are intimately related to the values of "liberal" society as a whole. Thus in the often-mentioned "conflict" between science and other social values, "it is not really science and morality that are in conflict but the morality of science and the morality of ordinary behavior."[34] This conflict exists, of course, and cannot be explained away by mere good will. It has all the more reason for existing, indeed, because science does arise in a morality of its own and not in a set of expedient principles that might bend more easily in the face of opposition from other social forces. One of the social consequences of science, we have seen, is that its critical rationality challenges the traditional moralities of other social activities. In part, this consequence is unintended, a direct expression of scientific morality, which must be expressed if it is to persist at all. This is the nature of morality. This unintended conflict between the morality of science and other social moralities is an intrinsic feature of our society.

But this conflict is in part avoidable. Too many scientists, and some of their lay brethren perhaps even more than they, have not acknowledged the importance of other values in society. They have been so much of the victims of a certain positivistic bias, indeed, that they have even denied that science itself rests on values. This bias has sometimes made them incorrectly assume that the method of science was an all-sufficient, exclusive form of human adjustment. We see more easily nowadays than we did in the late nineteenth century that this is not true. We see, as we have tried to show in Chapter Three, that society as a whole rests on a set of moral values and that science always functions within the context of those values. These social values pose certain non-empirical problems to which science, being concerned only for the empirical, cannot give answers —problems of meaning and evil and justice and salvation. The enthusiasm of science is a characteristic common in some measure to all moralities in the face of competing moralities. This fact being understood, we can avoid some, if not all, of the moral conflict that results.

B. The Social Responsibilities of Science

BECAUSE OF OUR increasing awareness of the social consequences of science, there has occurred on all sides recently an enlargement and intensification of concern for what are often spoken of as the "social responsibilities" of science. Nowhere has this increased concern been more manifest than among the scientists themselves, especially, of course, among the nuclear physicists, who see their own close connection with the atom bomb most clearly. Lee DuBridge, whom we mentioned earlier as President of the California Institute of Technology and Director of the Radiation Laboratory at M.I.T. during the war, and who is therefore a scientist very widely acquainted among his colleagues, has recently spoken of their change of attitude. "The net result of the war, I think," he says, "was that scientists are today somewhat more willing to play their part as citizens than they were before the war." Within this general change of attitude, however, the specific reactions of different scientists have varied somewhat. There have been at least three typical positions taken, none of them quite satisfactory as an analysis of the social responsibilities of science, but each of them revealing some characteristics of science which should be included in a more adequate statement.

One position taken by many scientists is that they have some general kind of social responsibility for the counsequences of their discoveries and inventions and that therefore they are immediately obligated to re-consider their position in society with a view to defining this social responsibility more precisely. This position is represented in its most organized and active form by an organization we mentioned earlier, in Chapter Five, the Federation of Atomic Scientists. This position is also taken by the scientists who read and edit *The Bulletin of the Atomic Scientists,* many of whom are members of the F.A.S. The scientists who take this stand have very zealously and very usefully tried to make clear to the general public the significance of new atomic energy possibilities, for example, the H-bomb. Their comments and warnings have been widely published in the newspapers as authoritative statements by scientists in a special position to enlighten the public.

Another reaction is one which quite explicitly accepts total responsibility for the social consequences of science and tries to prevent some of the most abhorred ones. There seem to be very few scientists who have taken this extreme view of their moral obligations to society. The most notable example has been Professor Norbert Wiener, the famous mathematician of M.I.T., who publicly announced his intention not to publish any future work "which may do damage in the hands of irresponsible militarists."[35] Many scientists, also publicly, criticized Wiener for this stand, saying that his action was altogether unrealistic even though his intentions were entirely good.[36] Wiener's critics correctly pointed out that to achieve his purpose he would have to stop his scientific work entirely, since he could not possibly foresee what its use might be. Actually his work has been, indirectly and directly too, of considerable use to the American military.

A third reaction expresses resentment both of the unwarranted acceptance of too much social responsibility by the scientists themselves and of the imposition of such responsibility on scientists by laymen. Professor Percy Bridgman, winner in 1946 of the Nobel Prize in Physics, who has had a long-standing concern for the social responsibility of science, was the most distinguished scientist taking this position. In a general article on the subject he speaks bitterly of "the legend of the responsibility of the scientists for the uses which society makes of their discoveries," and he sharply advises scientists not to accept "the careless imposition of responsibility, an acceptance which to my mind smacks too much of appeasement and lack of self-respect."[37] Professor Bridgman is no inhabitant of the ivory tower and therefore his words carry great moral force among scientists. That he was not one to shirk moral responsibility in science he had shown beyond a shadow of doubt when he issued his "Manifesto" in 1939, barring all scientists of totalitarian states from his laboratory, where he was making discoveries that would have been of direct use to the military forces of those countries. His has been no merely passive interest in science and its place in a "liberal" society.

What, then, is the social responsibility of science? Should a scientist feel his moral obligations discharged by any one of these three positions? Before we can try to answer these questions we need to recall some of the characteristics of science which we have already discussed.

We have to recall first what we have just said, that the social consequences of science are inevitable, that because of the uniquely strong place which science has in our society it will constantly interact with other parts of the society for both good and bad. In short, it is sometimes very difficult to live with science, as well as delightful, and we can only do so by learning to cope with its social consequences somewhat more adequately. This problem, we have seen, is a "social problem," a matter of social arrangements and social values, and it is not capable of solution in any degree by the natural sciences as such. It is the "social problem" which poses the question of the social responsibilities of science.

The second characteristic of science which we have to keep in mind is that we cannot predict, on the whole and especially over the long run, the particular social consequences which some scientific discovery will have. For example, atomic science directly depends upon the discovery by Röntgen of X-Rays, and yet no one could predict from his researches, occurring about 1900, what their present-day significance would be for the atom bomb. Or, to take a "good" consequence of his discovery, for cancer therapy. The more fundamental the scientific discovery, the greater number of direct and indirect consequences it is likely to have and the more difficult to predict its multiple good and bad applications. Examples like X-Rays can be enumerated at length from the history of science. It is also true that even minor discoveries have eventual convergent effects which could not be predicted when they were made. Science is a cumulative structure to which each researcher adds his little bit, the total often being synthesized in ways and used in ways which no single individual scientist could possibly have foreseen. Professor Wiener's position ignores this essential characteristic of science.

And finally we have to remember that science does not have its social consequences at a distance, in some kind of social vacuum, but rather constantly interacts with the rest of society to produce these consequences. To take the most obvious general cases, science is differently used, we know, by one or another government or political party, is differently used in war and in peace, and is differently used in prosperity and in economic depression. Social factors are highly variable too, and they interact with science all the time, thus rendering unrealistic the attribution to science alone of effects which have had multiple causes over a long span of time.

All this will, perhaps, make it clear that neither scientists taken as a whole group nor any individual scientist alone can be considered responsible, in any sensibly direct fashion, for the social consequences of their activities. The very specialization and interdependence of the parts of our society implicate every one of us in these social consequences. We are all, for example, more or less directly involved in the responsibility of war, if such can at all be thought to be a useful way of looking at things. The connection of the nuclear scientists only seems more direct and obvious nowadays than that of some other groups in the society. Science can be given no exclusive responsibility, that is to say, for the social and political problems for which all members of the society must take some measure of responsibility. The social consequences of science, so-called misleadingly, we have now seen, are social and political problems that can only be managed by the social and political process, to the extent that they can be managed at all. Even if they wished to do so, scientists could not be allowed to pre-empt the social and political function in society. For as scientists they are no more, and sometimes no less, competent in this function than other men. Certainly they are not experts in it by training and very seldom by experience. Clemenceau once remarked that war was much too important to be left to the military. In the same fashion, science and its consequences are much too important to be left to the scientists. In both cases, the instruments are much too important to our social purposes to be left wholly to the experts in using those instruments. They are the concern of all who have the responsibility for our social purposes.

On this view, if we may look at the case of war a little more closely still, some of us over-emphasize the importance of science in the conduct of war and in the prevention of war. Although science has changed the techniques of war continually during the history of man as much as it has changed all our other social techniques, war is a social reality all apart from the particular kinds of science it uses. War was evil long before the invention of poison gas, Haldane pointed out just after the first World War when men were still debating the morality of the new techniques which science had provided for that war.[38] Similarly, war was evil long before the atom bomb was devised, and it would still be evil if the atom bomb were outlawed. The British scientist, Eric Ashby, has advised his colleagues wisely and a little bitterly in this connection. "The prevention of war," he says, "is an urgent practical problem to be solved (if it can be solved at all) by political techniques, not by electronic Don Quixotes."

If we are agreed upon this analysis of the social characteristics of science, if we see its place in the whole of society, then we can re-phrase the question of which we are here speaking. We shall no longer ask, what is the social responsibility of science? We ask, instead, what contribution can scientists make to the social and political process of society. Or we ask, more generally still, what is the responsibility of a citizen with highly specialized and esoteric knowledge to his "liberal" society? Because increasingly some members of our society do have specialized experience and knowledge, the problem of the responsibility of the scientist is part of the general problem of the responsibility of the expert in "liberal" industrial society.

In a democratic society like the United States, of course, each individual scientist must choose for himself just what kind of responsibility he will assume for his membership in the scientific community. It is of the very nature of our society that social responsibility is largely a matter of moral obligation voluntarily assumed, and this holds for all of us, scientists and non-scientists alike. Our democratic values permit a great deal of exhortation to responsibility, but only a little compulsion. Now some individual scientists, like some other individuals, will not and do not feel morally obligated to participate actively in the political process. They are then subject, of course, to the moral judgment of their fellow citizens. This does not mean, however, that democratic moral judgment should or will always condemn the socially inactive scientist. For over quite a wide range of behavior we do acknowledge that some of our fellows may be called by other compelling interests, by other values, than direct political participation. We do grant a great deal, that is, to the man who cares overwhelmingly for his work, particularly when we admire what he is doing. It would certainly be unfair not to grant this privilege to some, at least, of our scientists, since we grant it to other kinds of experts and specialists. Here again we have to note that the scientist has no peculiar or exclusive social responsibility.

Furthermore, even when he does wish to participate actively in social affairs other than his scientific ones, the scientist may fairly claim the democratic privilege of choosing the kind of course which is most congenial to him and in which he thinks he can be most effective. Only a few scientists, for example, by the very conditions of their occupational specialization, can make a large contribution to direct political and social action. Yet some have done just this, at least for a little while in times of social crisis. We have seen how the "scientist-statesmen" of World War II, men like Conant, Bush, and Compton, took on a great deal of responsibility in the Government's use of science. In such direct political participation, the scientist deals with social problems and helps to form social decisions, bringing to the process his expert view of science both as a body of specialized knowledge and as a social organization with particular characteristics. Such direct political responsibility, however few the scientists who can assume it, is of great importance to American society.

The talent for such large and direct social responsibility is, unfortunately, no more common among scientists than it is among other specialist groups in the United States. Most scientists are limited to something much less than this. One of the more limited kinds of contributions scientists can make we have already mentioned. Scientists can do what the editors of the *Bulletin of the Atomic Scientists* do, that is, study their subject with a view to showing some of its possible social implications and keep the general public informed of these matters. Because of the great authority with which he is invested, the scientist can often communicate the meaning of new discoveries better than anyone else could. All such scientific dicta, however, should not over-reach the limits of the scientist's technical competence. The physicist, Louis Ridenour, has seen the importance of this responsibility we are here speaking of. "It is necessary today," he says, "to educate the non-scientific public to the Promethean nature of atomic energy and the true character of science. This education must be done so that all the people can participate in the decisions they will have to make concerning the organization of society in such a form that wars become less likely."[39]

Among these few alternatives, then, of no direct action at all or more or less limited action, each scientist must choose his course for himself, considering his own temperament, need, and competence. Professor Bridgman is only asserting an essential democratic right of all American citizens when he says society should not "insist on its right to the indiscriminate concern of all scientists with this problem."[40]

In the acceptance of any kind of social responsibility for science, two extremist positions should be rejected because of the dangers they involve for science. One is Professor Wiener's position, a kind of acceptance of exclusive responsibility. The danger here is that laymen may take scientists at their word, become convinced of the evils of science, and then hamper or even stifle science in the conviction that thereby they are only protecting society as a whole. Scientists who understand the limited nature of their responsibility will avoid the possibility of this boomerang effect. Another extremist position which may provoke unhappy effects upon science is the "ivory tower" position, which holds that scientists are interested only in "pure science" and are not at all concerned for the social consequences of their discoveries. The danger of this attitude is that society may come to think of scientists as a group of irresponsibles against whom it needs to protect itself. Men like Professor Bridgman, rejecting the extremism of the exclusive responsibility position, have to be careful they are not pushed into this opposite extreme. Fortunately, nowadays neither one of these extreme positions is taken by many scientists.

Perhaps one further, larger responsibility of science may be suggested finally. Science has the obligation, as we see it, to extend its method to the study of the social and political process itself. Concretely this requires that natural scientists develop at least a sympathy for the development of social science. As we shall see in our next chapter, there is no intrinsic reason why social science cannot exist as much as natural science. Indeed it has already advanced in our society farther than it ever has before in human history. Yet in testimony before Congress on the inclusion of the social sciences in the National Science Foundation, some important leaders among American natural scientists showed little sympathy for social science as it now exists and sometimes even little conviction of its real possibility. The fundamental values of their own scientific activity and of the society which supports it, we have seen in Chapters Three and Four, seem at the very least to require them to have a faith in the possibility of social science. At this point in the history of science it belongs to the social responsibility of all scientists to support scientific analysis of the social and political problems which so dangerously threaten the existence of science itself.

C. *Can Science Be Planned?*

THE SOCIAL CONSEQUENCES of science in the modern world have inevitably raised the problem of "planning" science, of controlling it in such a way as to maximize its favorable effects and minimize the harm it can do. The "scientific humanists," for example, of whom we have just spoken, not only oppose what they call "the frustration of science" but they want to "plan" science, they say, for the greater good of society. In England especially their books and speeches were very influential throughout the 1930's and their point of view seemed to be sweeping the field clear of any opposition at all to "planning" in science. But in 1940, finally, inspired chiefly by Professor Michael Polanyi and by Dr. J. R. Baker, another group of British scientists, who had been unhappy all the while, formed the Society for Freedom in Science, in specific opposition to the views of Professor J. D. Bernal and his colleagues in the camp of the "scientific humanists." By June, 1946, the Society had a membership of more than 450, with 250 in Great Britain, 176 in the United States, and the rest scattered throughout the world. In the United States, Professor Percy Bridgman, whose concern for the social problems of science we have now several times seen, became the unofficial leader of the Society.[41]

This conflict over "planning" among the scientists themselves is in part an aspect of the larger conflict in our society over social planning. We live in a time of great change and in a time, therefore, when men demand more social control in human affairs generally. The larger social problem, of course, has become the focus of great political and ideological dispute, the focus of opposition between "right" and "left," "liberal" and "conservative," "socialist" and "capitalist." In this atmosphere, it is not strange that the problem of "planning" science has become involved in the larger problem and that scientists have often taken stands in the larger political and ideological terms.

Now the two problems, the greater and the lesser ones, are indeed related in many ways and should be considered together for many purposes. We have seen that science and society are vitally connected throughout a whole range of their workings, and so we know that "planning" in one area will necessarily affect other parts of society. Yet there are also some important differences here, some matters more restricted to the nature of science. We have seen that science has its own characteristic set of values, its own special types of social organization, and its own peculiar processes of discovery and invention. The problem of "planning" science, therefore, whatever its more general connections, needs to be considered in the light of these special characteristics of science. Because it is a "social problem," furthermore, we shall not be surprised to find that the word "planning" has been used to mean many different things actually, not one thing. It will be profitable for us to separate these different meanings and consider each one in the light of some of the analysis of the social aspects of science that we have been developing up to this point. We can do this fairly briefly, without repeating what we have said in detail. And perhaps this approach will show that there is less difference between the partisans and the opponents of "planning" than they themselves sometimes think, that they have a large agreement on important concrete problems of scientific organization all the while that they enter into conflict on the more abstract matter of "planning" as a whole.

We can see this very clearly if we start by taking "planning" in its simplest, everyday sense, the sense of setting oneself fairly specific goals and of doing the best one can to devise techniques for achieving those goals. There is very little conflict about this kind of "planning" in science, this attempt to be as rational and efficient as one can in reaching established goals, especially those about which everyone is agreed. There is little disagreement over such matters as the planning of careers by individual scientists, planning by university science departments to "cover" all the specialties of their field and to choose their members accordingly, planning by industrial enterprises to expand research facilities to increase their profit, and planning by Government to use science for its multifarious and recognized social responsibilities. As the results show, at its best this kind of planning in science is very successful, proceeding on the basis of the wisdom of scientist-administrators with long experience of the special nature and needs of science. All that we can hope for here is some improvement in the knowledge we have of science so that we can be a little more skillful, a little more efficient in achieving the goals we are all agreed upon. Professor Bernal is speaking of this kind of "planning" and this possibility of improvement in our knowledge when he says, "It does not, of course, follow that any kind of organization would be appropriate for science. The mere task of finding the kind of organization needed for science is itself a scientific problem."[42]

There is a great deal more conflict, however, when "planning" includes the question of the goals of science, not merely or primarily the means to generally approved goals. We have seen that science as a whole has different and sometimes conflicting goals. For "pure" scientists, the essential goal for scientific activity is the extension and improvement of existing conceptual schemes. For other scientists and for many non-scientists, the essential goal is the successful application of existing scientific theories to the practical purposes of the industrial, military, or governmental organizations of the society. In short, there is both "pure" and "applied" science, each necessary and each socially legitimate. But "planning" in science is sometimes intended and sometimes taken to mean that only "applied" science will be legitimate, that "pure" science will be abandoned. Probably no one has ever actually proposed such an extreme course for "planning" in science, yet this is what some of its more ardent exponents have seemed, to some of its more ardent defenders, to suggest. The Society for Freedom in Science seems to think this is what the "scientific humanists" have in mind, for the Society has set down as the very first of its five fundamental propositions about the essential nature of science that, "The increase of knowledge by scientific research of all kinds and the maintenance and spread of scientific culture have an independent and primary human value."[43] Yet Professor Bernal, for all his talk of "planning," had said long before, "Throughout any plan of scientific advance it would be necessary to keep a just proportion between fundamental and applied research and to maintain at all times the closest contact between them."[44] We have already seen in Chapter Four that "pure" and "applied" science are both necessary and also that they are necessarily interconnected. On this matter all who understand the possibilities of social control in science are agreed.

The disagreement in this matter, then, reduces itself to what is still a genuine problem, that of apportioning the scarce social resources of science in some sensible fashion between the two goals. Now it is not easy to be "sensible" here, because scientific resources are not completely flexible, any more than other kinds of resources, and because we do not know as much as we need to for this kind of "planning." A great deal of it goes on anyway, usually by informal rather than formal means, but of course in wartime and in other social crises there is a great increase in formal "planning" looking toward *more* "applied" science and *less* "pure." It is inevitable that men will differ in their social purposes somewhat, and this basic ineradicable residue of difference is the source of conflict in the choice of *how much* "pure" science and *how much* "applied" science we are to have. Yet practical compromise can be much more readily achieved if the area of conflict is thus narrowed and seen to be legitimate within the more general agreement on the necessity for both kinds of science.

The competition among alternative social purposes, the necessity for choice among several desirable goals, is a fundamental feature of social life. Science as an end in itself cannot hope to evade this competition any more than other social activities can, and in this sense it must inevitably be somewhat "planned." On this understanding, it can only compete more or less successfully with other social purposes by making clear the different functions of "pure" and "applied" science and by making clear the conditions under which each can be successful. Instead of resisting all competition with other social activities and all allocation of its resources as "planning," science should seek to maximize the relative achievement of its own purposes without denying the significance of other social goals. The two things have always gone together and can continue to go together without the unlimited conflict which some discussions of "planning" in science take as a basic premise.

Another meaning of the term "planning" which is often used is that of the ability to predict the course of scientific discovery. Professor Polanyi, for example, says, "And here indeed emerges the decisive reason for individualism in the cultivation of science. No committee of scientists, however distinguished, could forecast the further progress of science except for the routine extensions of the existing system. The problems allocated by it would therefore be of no real scientific value."[45] Now we have seen what difficulties lie in the way of successful prediction in science, yet Professor Polanyi seems a little pessimistic about its possibilities. After all, university, Government, industrial, and private foundation grants committees are every day making some rough predictions about the likely course of scientific discovery. They do not always follow the main chance, of course, yet their distribution of funds represents at least a balance between the probability of success and other interests in pursuing a certain path. Perhaps, in contrast to Professor Polanyi, the "scientific humanists" seem to be a little optimistic about our ability to predict where scientific progress will be made, yet here again there is, at the concrete level, a great deal of agreement about the nature of science. "Science," says Professor Bernal, "is a discovery of the unknown, is in its very essence unforeseeable."[46] Although he says we do not know just what we will find in any next step in science, "we must," he says, "in the first place, know where to look. Some amount of short-range planning has always been inherent in scientific research." Professor Bernal wants more self-consciousness about what already exists, namely, this short-range prediction and planning, and probably also he wants to extend prediction wherever possible. Since this is what already occurs in science, there is not here any grounds for irreducible conflict about this aspect of "planning" in science. When stated in appropriately concrete terms by groups of scientific specialists, a certain amount of forecasting, as we have seen, is entirely possible in science. Partisans of "planning" in general and partisans of "freedom" in general do not differ so much as they think and could probably very profitably collaborate on the kind of concrete prediction and programming that goes on every day in science.

On this point, incidentally, even the Russians are not deluded. "There is no possibility, of course, of planning out 'unexpected' scientific results and discoveries," says the Russian scientist, Vavilov, President of the Academy of Sciences, "but all true science must contain a very large proportion of well-founded anticipation and prevision." For instance, he says, "our contemporary knowledge of the structure of the atom nucleus allows us to plan out for many years to come, with a large degree of confidence, much of the theoretical and experimental work to be done in this field."[47]

And finally we come to one last meaning of "planning" where once again the controversy between those "pro" and those "con" has been carried on in too general terms. This meaning of "planning" has to do with the appropriate social organization of science. We have seen in Chapter Four and elsewhere that there are different kinds of social organization in science, some of it highly formal and bureaucratic in type, some of it more informal. We have seen that although the amount of formally organized scientific work is increasing, science as a whole must necessarily remain only informally coordinated and controlled. What Professor Polanyi and his fellows in the Society for the Freedom of Science fear is the prospect that science as a whole will be bureaucratically, monolithically organized under some external political authority. That is why Professor Polanyi argues so ably and so eloquently for control of science by what he calls "spontaneous coordination," by which he means what we have described as informal organization, and which, we have seen, exercises a definite pattern of beneficent authority in science.[48] But Professor Bernal and the "planners" in science seem at least open to this possibility, since, as we have already quoted him, Bernal says that not any kind of organization "would be appropriate for science. The mere task of finding the kind of organization needed for science is itself a scientific problem." This is certainly a concrete enough task on which all scientists in a "liberal" society could profitably unite. Indeed, the whole problem of "planning" in science would be freed of much acrimony, and our understanding would be considerably advanced, if more attention were paid to the area of actual agreement among scientists and less to their general and ideological differences, more to the concrete matters of scientific activity and less to dispute about ill-defined terms like "planning." In short, more of the method of science itself is what we need here.

XI

The Nature and Prospects of the Social Sciences

WE HAVE up to now been dealing with the social aspects of the so-called "natural sciences," that is, with the physical and the biological sciences, paying only slight and incidental attention to the so-called "social sciences." We may now turn, finally, directly to the consideration of this latter group of sciences. This procedure may seem only "natural" to many, who consider that there is indeed some essential difference between the natural and the social sciences. But it is surely clear by now that we hold an entirely different view of the matter. Everything we have said already has been based on the assumption that social science is not only possible but even essentially the same as natural science. The empirical facts with which we have been dealing, the facts about the social organization and the social relations of science, are as much subject to scientific investigation as any other class of empirical phenomena. Science is a unity, whatever the class of empirical materials to which it is applied, and therefore natural and social science belong together in principle.

A great deal of what we have already said holds equally for social and natural science, in respect of their rational method, their supporting values, their modes of social organization, their consequences, and their social control. But though they are essentially the same in principle, natural and social science in the modern world are obviously at significantly different stages of development and acceptance. This disparity of development, and not any dissimilarity of its fundamental character, is the justification for treating social science separately. And because so much of what we have said about natural science holds equally for social science, we shall speak of some of the social aspects of social science in this chapter in a relatively abbreviated fashion, concentrating on the special problems of its present condition which arise out of its less advanced state of development.

There is no need to specify any more closely for social science than we have for natural science just which activities ought to be included in it, which excluded. Science as a whole is fuzzy around the edges, blending into common sense and everyday practical activity. And the individual sciences overlap with and trickle into one another in unexpected and fruitful ways. All this is true for social science, and so we shall use that term henceforth to refer *roughly* to a group of academic disciplines—and their practical applications—which are usually called economics, political science, psychology, sociology, and anthropology. A great deal that is called "history" actually works toward the same ends as these five disciplines we have named. One essential characteristic of all these social sciences is that they deal with the social relations between human beings, that is, with those relationships between human beings in which they interact with one another not as physical objects merely but on the basis of mutually attributed meanings. This capsule definition is not intended to be a complete or satisfactory one, but only a first approximation to the necessity of defining the class of empirical facts which are relevant to the conceptual schemes of the social sciences. The only fully satisfactory definition of a science, we saw in Chapter One, is a complete statement of the substantive theories of its conceptual scheme. It is not necessary for our present purposes even to attempt this task for the social sciences.

We may repeat that the five academic disciplines we have just named do not exhaust the list of those which are in any way concerned with what we have seen in Chapter One is the primary task of any activity that pretends to be a science: the construction of ever more abstract, ever more generalized, and ever more systematic conceptual schemes. For instance, certain parts of jurisprudence, a great deal of the work done by historians, aspects of applied professions like medical psychiatry—all these, and other studies, too, contribute directly and indirectly to the essential purpose of social science. But we shall keep in mind primarily the five we have selected because their *main* effort and aspiration is to develop conceptual schemes and thereby to become independent social sciences. Our five disciplines—economics, political science, psychology, sociology, and anthropology—afford us full opportunity to discuss the social aspects of the social sciences insofar as their lesser advancement makes them different from the natural sciences. What we say of these five may be taken to apply equally to those parts of any other academic disciplines or everyday activities which have the same goal of becoming social sciences in the sense here used.

Before proceeding to our discussion of the possibility for still further developing conceptual schemes in the social sciences, a word should be said about something that is important in the relations between the social and the natural sciences. It is very much worth remarking, that is, that these two rough classes of science are not completely separable from one another, although it is very commonly assumed that they are. The conceptual schemes of the two classes of science will be different, it is true, because the defining characteristics of the empirical phenomena in which each is interested are different. This difference of conceptual schemes holds also, of course, *within* the natural sciences—between the physical and the biological sciences, for example. But in practice, in the actual investigation of specific and concrete behavioral problems, the social and natural sciences overlap. Just as biology and chemistry now overlap and cooperate in the study of biochemistry, because certain concrete phenomena require it, so too the natural and social sciences overlap and must also cooperate. There are several examples of already existent cooperation. For instance, there is the now flourishing study of psychosomatic medicine, the very name of which indicates the concrete inseparability, in some measure at least, of the natural and social sciences. It is the fundamental premise of psychosomatic medicine that the meaningful aspect of human behavior is in direct interaction with its physical and its biological aspects. The effects of anxiety and other psychological conditions are known to manifest themselves in a whole range of physiological symptoms, such as peptic ulcer, arthritis, and allergies. It follows from this interconnection of the sciences that progress for each kind, for natural as well as for social, depends in some areas on the progress of the other. Scientists as well as the laymen who support the sciences should consider the significance of this fact, that science is more a unity than they sometimes think.

Let us look at this matter a little further. There are other practical social problems where the individual contributions made by the natural and the social sciences are somewhat more easily separated than in the case we have just mentioned, but where their very close collaboration is mutually fruitful and even necessary for any practical success. In the applied science of industral and management engineering, for example, the social sciences of psychology and sociology are now recognized to be as essential as mechanical engineering itself. The "rationalization of industry" movement may have started with engineers like F. W. Taylor, Gantt, and Frank Gilbreth (who were more social scientists, indeed, than they knew), but it has moved on in its development to include among its leading contributors social scientists like Elton Mayo and F. J. Roethlisberger.[1] The famous studies by these latter two and others at the Hawthorne Plant of the Western Electric Company, in which they demonstrated that the "social factor" was an essential component of the industrial situation, started in the attempts of the lighting engineers, it may be remembered, to trace the effects of illumination on work efficiency. Natural science found it had to include social science. This was also true during the recent war. For instance, a psychologist who worked closely with natural scientists on the design of air-borne instruments during the war says, "Experimental studies of the efficiency of air-borne weapons could not be made successfully without the cooperation of mathematicians, physicists, engineers, and psychologists."[2] Or, to take one last example of this overlap of the social and natural sciences, it has recently become very clear that any really satisfactory studies in demography, or population problems as this field is often called, will find the social sciences as indispensable as the biological sciences. Birth and death rates involve the close inter-working of social factors and biological factors.[3]

We may now return to what is the basic problem, the question of the possibility of developing the conceptual schemes of social science to a point more nearly equal to those of the natural sciences. It is certainly obvious to everyone, to the layman as well as to the informed specialist, that the social sciences now have a much lower degree of power and autonomy than do the natural sciences. The informed specialist knows this because he sees the absence of highly developed, empirically tested conceptual schemes in social science. The layman knows it because of the absence of that wide range of practical applications in his daily life which so much impresses him in the case of the natural sciences. But although everyone is convinced of this relative weakness of the social sciences, some specialists and some laymen go a great deal farther still in their views. There are some, that is, who even feel that social science is more than just temporarily weak, it is impossible. There are some who are convinced that human behavior is inherently irregular, capricious, and indeterminate and that therefore social science is *in principle* a chimera which only the foolish waste their time pursuing.

If our view is the opposite one, that social behavior is determinate and that therefore a highly developed social science is possible, how do we justify it? Let us start with first things, as we did in Chapter One when we were speaking more directly of the social origins of rationality and natural science. Not only are social phenomena just as much empirical matters as are the objects of natural science study, but in all societies there is considerable rational knowledge of empirical social phenomena. Indeed, we may put it as strongly as we did for the case of natural science: human society is impossible without considerable knowledge of this kind and its accompanying social technology. In all societies there is at least an embryonic science of society. To take only the roughest and most general examples, in all societies men know how to train the younger generation to carry on the essential social tasks; men know how to order their daily routine and their emergency affairs; they know how to govern; in short, they know how, *at least tolerably well,* to make their social affairs orderly, predictable and stable. They do not, of course, have this knowledge in any highly abstract, generalized, or systematic fashion. They do not, that is, yet have the conceptual schemes of an advanced social science. But because they do not now have such a social science does not mean in principle that they cannot have it. On the contrary, insofar as it is always possible to take common-sense rational knowledge and improve it in the scientific direction, then social science is in principle possible because it already exists in at least a relatively undeveloped state.

This is only a minimum claim, however, this statement that social science is in principle possible. What has been done with this possibility? Granted that social science in the modern world is much retarded in comparison to natural science. But perhaps this is not the only comparison we should make. Perhaps we shall have a better view of the real possibility of social science if we compare its development in modern Western society with its development elsewhere, in other modern "civilized" societies as well as in "primitive" societies. In this societal perspective, it is readily apparent, social science has reached a much greater stage of advancement in our society than it ever has anywhere else. Perhaps a satisfactory social science will appear to us so much the more likely if we look occasionally at whence present social science has come, at how much better it is than what is available in other societies, rather than if we only concentrate on how far it has to go to catch up to natural science in our own society. We will also do well to remember, in this connection, what enormous progress has been made during the last one hundred years by the biological sciences and their chief direct beneficiary, the medical sciences.

This more hopeful view of developing a mature social science does not explain, of course, how it has happened that natural science has outdistanced social science in our society. It does suggest, though, because both kinds of science have developed more rapidly in our society than they have elsewhere, that there is something characteristic of modern Western society which is favorable to science of all kinds—physical, biological, and social alike. This "something characteristic" is the set of social values and social conditions which, as we showed in Chapter Three, is an especially favorable one for the development of science. Favorable for the development of *all* science, we hold, social as well as natural. Still, we do not know very well why social science has lagged behind natural science in its development, any more than we yet understand precisely why natural science itself did not mature sooner in human society, or why the biological sciences are less advanced than the physical sciences.[4] Our social science is not yet good enough to treat satisfactorily of either of these problems. All we can say is what we have said in Chapter Two in our crude explanation of the emergence of a highly developed natural science in the sixteenth and seventeenth centuries: that directly and indirectly a great many different social factors affected the emergence of this complex set of social activities, social values, and theories we call by the simple name of "science." Any attempt to account for the relative lack of development of the social sciences must go beyond some simplistic formula—for example, that social science is just impossible—to some treatment of their inter-relations with several other social factors. Perhaps one useful way to commence this task is to study the present position and the prospects of the social sciences.

One of the best auguries for the successful development of a more mature social science in our society is the increasing awareness among social scientists of the nature of the scientific task which confronts them. More and more their self-conscious ambition is the same as that of the natural scientists, to create a set of highly determinate theories for the explanation of empirical social phenomena. Increasingly they are learning the significance and functions of abstract and systematic conceptual schemes. Increasingly they abandon merely logical and armchair speculation about social life and seek for instruments wherewith they can empirically test their theories against reliable social data. The self-conscious use of theory in empirical investigation is, to be sure, highly uneven over the field of the social sciences. But all five disciplines we have been speaking of, are, each in its own fashion, striving toward the construction of at least limited conceptual schemes; and they are all also seeking for new instruments to test these constructions. Some of these new instruments, like the interviewing, questionnaire, and polling techniques, have large promise for all the social sciences.[5] Indeed, there is now offered even a general conceptual scheme underlying all of the social sciences; that is, a general theory of social action has now been formulated with the explicit purpose of providing the most abstract, generalized, and systematic conceptual scheme which is now possible for the social sciences.[6] Whatever the fruitfulness of this scheme may be can only be judged over the long run when it has been put to the scientific test of further empirical verification. For the present, in any case, it serves as a model of the kind of theoretical sophistication which is now possible in the social sciences; and, perhaps more important still, it provides a focus for the inevitable formulation of further advances in the conceptual scheme which must sooner or later become the basis of a satisfactory social science.

In our discussion of the nature of science in Chapter One we noted the indeterminacy of "common sense" in comparison with systematic natural science and the way in which the latter affects and changes the content of what is held to be true by "common sense." It is perhaps in regard to social behavior that "common sense" is most strikingly indeterminate, being marked by all manner of inconsistencies and expressions of feeling rather than by precise and valid statements of social fact. If one wants to see how true this is, all he has to consider, for instance, is the many different things that are alleged about "human nature." This indeterminacy reflects the weakness of social science, which has not yet been able to affect "common sense" nearly so much as has natural science. Nevertheless, some influence exists and in some cases it is not even inconsiderable. Take, for example, the way in which the relatively sophisticated theory of Freudian psychology has in recent years infiltrated the thought and speech patterns of people throughout our society. Probably as the strength of social science grows, and as its accepted influence on "common sense" is greater, there will even enter into "common sense" a view that does not yet exist there, the view that social science, like natural science, is better than "common sense." Even what little there is in social science of genuine usefulness for practical social conduct still has to combat the "common sense" conviction that in social matters every man is his own best expert. Actually much "common sense" about social behavior is outmoded social theory; for example, a great deal of it nowadays is still outmoded Social Darwinism, a set of theories about the nature of man and society which is now demonstrably inadequate. The more adequate notions of social science make only slow headway against the inertia of "common sense" conviction and opinion.

The prospect for social science is in another respect very favorable in our society. What we said in Chapter Three about the general congruence of the social values and social organization of "liberal" society with the development and maintenance of natural science holds also for social science. The value we place upon "critcial rationality" and the structure of our occupational system, to select only two of the social conditions we have already discussed, are both favorable to social science. Not only does the occupational system, for example, make it possible to have a group of specialized social scientists working full time at their task, but in turn it requires considerable social technology for its efficient operation. This is an important stimulus for social science advance. As we have just said, the "rationalization of industry" is dependent on economics, sociology and psychology, as well as on improvements in machine technology. We may, indeed, wonder whether social science would have achieved even its present relatively low state of development in anything but a society which places so great a value upon "critical rationality." Our society is unique, we have seen in our discussion of the social consequences of science, in the extent to which what Max Weber called "the process of rationalization" has gone. This process includes and is based upon a critical examination of all social organization, all social values; and this is an undertaking which no other society has ever tolerated as we have. It is out of this freedom to investigate rationally the very fundamentals of society that social science has come. This freedom is also a warranty for its future achievements.

The social organization, too, of social science is in general like that we have described for natural science. Social scientists carry on their activities in universities and colleges, in industry, and in the Government, although there are fewer social than natural scientists in industry. Professional social scientists have only been in existence since very late in the nineteenth century; like natural science, social science was a subject for "amateurs" for a long time before it became an occupation for professionals. The following figures from the National Roster of Scientific Personnel indicate approximately the number of professional social scientists:

Anthropology	683
Economics	7,349
Political Science and Public Administration	2,742
Psychology	6,985
Sociology	2,729[7]

The cultural values that control the behavior of social scientists are also the same as those that we have seen, in Chapter Four, to be important for their natural science colleagues; and they have the same overlap with the values of the larger "liberal" society and also the same divergences, for example, with respect to what we have called the value of "communalism" in scientific property. Social scientists are perhaps less able in practice to realize some of the scientists' values, say, the value of emotional neutrality; but if this is so, it is not so much because the value is any less strong among them as because their theories and their research techniques are not yet powerful enough to provide the safeguards which are built into the more developed equipment of the natural sciences. One of the more important functions of a highly developed conceptual scheme is to structure scientific research in such a way as to eliminate certain kinds of errors and certain lapses from the scientific norms.

In the matter of the public evaluations of the role of the social scientist, the situation is somewhat more obscure than is the case for natural scientists. We have seen in the poll data from the North and Hatt study of popular evaluations of occupational status that the social scientists are ranked high, in the upper group of occupations, along with the natural scientists. But this apparent equality with the natural scientists, this general popular esteem as expressed in a poll, does not seem to exist in all sections of the American public. Perhaps the high ranking for the occupation of social scientist in the North and Hatt poll occurs because it is a relatively vague one to many people and because in that situation it is assimilated to the approved status of "professor" or "scientist" in general. Expressions of evaluation of social science by certain influential groups in American society actually cover a wide range of approval and disapproval. Because of the effects on the development of social science that these groups can have, it will be useful for us to look a little more closely at some of these different attitudes.

There is, first of all, a group of people of influence and prominence in natural science, in education, and in public affairs who assign a relatively low status to social science. At least this is what must be inferred from what they say in public, for instance, at the Congressional hearings in 1946 on the proposal to include the social sciences in the National Science Foundation. We have already seen that the attitude of these people toward social science has carried the day and that social science has not been included in the National Science Foundation. The views of this group, as expressed at the hearings, have been summarized as follows in a critical article by the sociologist, George Lundberg: "(1) Man and his behavior are not a part of nature that can be studied as basic, 'pure,' natural science. Social science, therefore, is a non-descript category consisting mainly of reformist and propagandist ideologies and isms. (2) The methods of social sciences are so widely at variance with those of other sciences as to make it inadvisable to attempt to administer research in the social science under the same organization (a) for fear of discrediting the other sciences, and (b) because people qualified to direct research in the other sciences would not be able to judge what constitutes valid or desirable social research. (3) Social research is especially in danger of falling a victim to pressure groups or of being corrupted by the government itself. And, finally, (4) After all, we know the solution of social problems through the historic pronouncements of seers and sages, past and contemporary, and all that is needed is more education to diffuse this lore and arouse moral fervor in its behalf."[8] We have already dealt with some of these criticisms, for instance, the notion that "common sense" is better than science, and we shall consider some of the other objections in a little while. For the moment we are concerned only with the expression of opinion.

These views of social science seem also to have some scope and force among the less well educated public. We have no direct evidence for what these other people think, but we may infer what some of them might say from what Congressman Brown of Ohio has said about the matter, also in the hearings on the National Science Foundation Act. The vigor and color of his phrasing suggests that his attitudes are a little more like those of the general public than of the academic groups we have heard from above. "Outside of myself," says Mr. Brown, "I think everyone else thinks he is a social scientist. I am sure that I am not, but I think everyone else seems to believe that he has some particular God-given right to decide what other people ought to do. The average American does not want some expert running around prying into his life and his personal affairs and deciding for him how he should live, and if the impression becomes prevalent in Congress that this legislation (for the National Science Foundation) is to establish some sort of an organization in which there would be a lot of short-haired women and long-haired men messing into everybody's personal affairs and lives, inquiring whether they love their wives or do not love them and so forth, you are not going to get your legislation."[9] We shall also speak later on of some of these matters raised by Congressman Brown, especially of the challenge which social science seems to offer to established values and social routines.

But although some very influential natural scientists have a negative attitude toward social scientists, there are a great many others who do not. Some evidence from a Fortune Magazine poll, for example, indicates that a very large number of natural scientists, the majority, indeed, are willing for the Government to support social science as well as natural science. To the question, "Do you think the social sciences should share in any disbursement of federal funds for research?", 81% of the natural scientists in the universities, 83% of those in Government, and 76% of those in industry reply "yes."[10] Perhaps even more significant on the favorable side, because of the scope of his influence both on scientists themselves and on the general educated public, is the view that President Conant of Harvard has recently expressed. It is also important to note that President Conant has only recently changed his mind to this view. "It is my belief," he says, "that methods have already been developed to a point where studies of society by competent scholars can provide basic information to assist all those practical men who struggle with the group of problems we list under the head of human relations."[11] And lastly, to take the most optimistic view of all expressed by a natural scientist about the possibility and prospects of social science, Julian Huxley says, "We need have no fear for the future of social science. It too will pass through similar phases from its present infancy. By the time that the profession of social science, pure and applied, includes as many men and women as are now engaged in natural science, it will have solved its major problems of new methods, and the results it has achieved will have altered the whole intellectual climate. As the barber-surgeon of the Middle Ages has given place to the medical man of today, with his elaborate scientific training, so the essentially amateur politician and administrator of today will have been replaced by a new type of professional man, with specialized scientific training. Life will go on against a background of social science."[12]

The position of social science in American society, these different expressions of attitude show, is not so clear-cut nor so widely approved as that of natural science. There remains some influential opinion which holds it in low esteem and opposes the extension of its domain. On the whole, however, there seems to be an increasing recognition not only of its real possibility but also of its necessity.

Social scientists, like natural scientists, may be grouped into two classes, those who are relatively more interested in "pure" science and those who are primarily concerned with applications of the conceptual schemes developed by the first group. There is considerable overlapping of the two groups, more so than there is in natural science because of the weaker condition of the conceptual schemes of social science. It is harder to concentrate attention on the development of theories when one is not quite sure of what is theory and what is sentiment. As is inevitable in science, however, the overlapping and the mutual interpenetration of "pure" and "applied" social science are beneficial to both. In the field of attitude research, for example, there has been a close and profitable connection between "pure" social psychology and its applications in public opinion polling for practical purposes.

The heart of "pure" social science research is to be found in the university, in the academic departments of the disciplines we have called "social science." It is no small advantage for modern social science that it is already established in the universities and does not have to work its way into them from the outside. The university, as we have seen in Chapter Six, is the main trustee and innovator of our cultural heritage, and its approval and control are important insurance that social science will have successful progress. Through affiliation with the university, social science is required to aspire to and maintain the values and standards of science in general. But despite its strong, essentially impregnable position in the university, social science is not accepted there wholly with favor. At least certain aspects of social science come under disapproval and sometimes attack from two established quarters within the university. On the one hand, some natural scientists do not admit the real possibility of social science, as we have seen from their objections to its inclusion in the National Science Foundation. And on the other hand, certain scholars in the humanities regard social science as a-moral and as a danger to the cherished values of American society.[13] We shall say more later about this problem of the relation between social science and values; it is a question which always arises in any serious discussion of the nature and prospects of the social sciences.

Perhaps the largest field for "applied" social science research, rather than "pure," has beeen in the activities of the Government. The regular and permanent employment of social scientists by the Government is a relatively new thing, but a great deal of "applied" social research has been carried on by the Government for quite a while now. Such research was usually for the purpose of discovering the social conditions on which some particular legislative proposal must rest. There has, for example, been a whole series of Congressionally-sponsored investigations and reports on broad social and economic problems. To go no farther back than the beginning of the twentieth century, Congress has empowered such outstanding researches of this kind as those made by the Industrial Commission of 1907, the National Monetary Commission of 1908, the Industrial Relations Commission of 1917, and the Joint Commission on Agriculture Inquiry of 1921.[14] And just to show the scope of this social research: the Industrial Commission of 1898 collected vast amounts of social data which it issued in 19 reports; the Immigration Commission published 42 different reports; and the National Monetary Commission published 23 reports. The reports of the National Monetary Commission, for example, led to the establishment of the Federal Reserve System. In addition to these more comprehensive reports, of course, Congressional committees have for a long time now been making special social surveys and studies of more limited scope.

In addition to the Congressionally-sponsored research activities of the Government, there has been a long series of important researches initiated by various agencies in the Executive Department. For example, there have been:—in the early 1930's, President Hoover's Commission on Recent Social Trends; the excellent studies, to some of which we have referred many times, of the National Resources Planning Board; the large-scale investigation of economic activities by the Temporary National Economic Committee; the three-volume report of the Attorney General's Patent Inquiry; and, only most recently, the Hoover Commission's study and recommendations on the problems of administrative organization in the Government. Social science research has been of no small importance to the Government in all these investigations.

The use of social scientists by the Government, however, passed a major turning point during the Depression of the 1930's. With the generally increased scope of Government activities which began in that period, a great many social scientists from the fields of economics, political science, and sociology entered the permanent employ of the Government. These first employees, some of whom have since left to go into university research, formed the basis of a continually expanding corps of Government social scientists. By far the largest part of this expansion occurred, as it did also for natural scientists, during the recent war. As an illustration of the present and possible future usefulness of "applied" social scientists in the Government, let us look at their war work in a little detail.[15]

Economists made up probably the largest group of social scientists doing research for the Government during the war. They worked in such programs as price control and rationing, taxation and war finance, war production, and manpower planning. There were also economists in the intelligence groups of the Armed Forces, in the Office of Strategic Services, and in the Foreign Economic Administration. Anthropologists were also useful to these intelligence groups, especially those who had a first-hand knowledge of actual or potential foreign combat areas. The largest task undertaken chiefly by anthropologists, however, was in the Foreign Morale Analysis Division of the O.W.I., where a study of Japanese society was made with a view to discovering the ways of influencing the morale of civilian groups and fighting units alike.[16] As for political scientists, a large number of them were employed by the Division of Administrative Management of the Bureau of the Budget to set up new administrative agencies and to make the old ones more efficient. Psychologists were used by both the Army and the Navy for the construction of personnel classification tests and for the selection and training of air pilots. In the Army Air Forces especially, psychiatrists and clinical psychologists were employed to maintain the proper psychological conditions for high fighting morale.[17] Sociologists and social psychologists in largest numbers were found in the Research Branch of the Information and Education Division of the Army. This group studied the attitudes and morale of soldiers in a wide variety of social situations: the use of their equipment, officer-enlisted man relations, Negro-white relations, the point system for discharge, and performance in combat.[18] Now these several examples of the work that was done by social scientists in the Government during the war do not exhaust what was actually accomplished, but they do indicate the kind of beginning that has been made in the application of social science—however undeveloped, relative to natural science, its conceptual schemes and instruments may be.[19]

The other main area for the application of social science knowledge and research methods is in American industry and business. In this area it is much harder to get information of any kind about the use of social science than about the use of natural science. This at least was the experience of the National Resources Planning Board in its report on Business Research, which was the third of its three volumes on American scientific research.[20] In its first two volumes, on research in Industry and in the Government, the Board's researchers were able to get a pretty good general and also detailed statistical picture of what was being done and what our scientific resources were. The best it could do for Business Research was to study the practice of thirty-three anonymous business firms, these being selected as well as possible from among those with the longest experience and from different fields of business. Five were manufacturers of industrial goods, thirteen of consumer goods, three were retail distributors, four were public utilities, and eight were service organizations.

A number of interesting things about the use of social science appear in this study. Perhaps the most important is that a great deal of limited social research is carried on which is either not recognized as such or not acknowledged by business men. Because they associate the term "research" only with the physical or chemical laboratory, many business men do not consider the collection of social data to be research. Social research in business, then, is usually called "business analysis" or sometimes "economic analysis." Within the last few years, however, somewhat more favorable attitudes toward self-conscious social research on its own terms has been appearing.[21] Social research in industry has entered in an unacknowledged fashion. "Most of the existing business research units," says the National Resources Board report, "were established subsequent to 1920 and originated by the association of an economist with a major executive officer."[22] By far the largest number of social scientists in business and industry still are economists, with applied psychologists doing personnel work perhaps next largest in number. The chief tasks for social science research in business are in personnel, scientific management, operational analysis, administrative organization, industrial relations, marketing and the analysis of social statistics, and Government relations. Many of these tasks run into one another, of course. The quality of the social research now done in business varies enormously among the companies which support it. Since there are almost no established standards, almost anything goes in some places. But the best of such research is very good indeed, and it has developed methods and collected data which would be very useful to basic social science. In this connection the National Resources report says, "In the files of business concerns there reside today valuable data, ingenious methods, and practical applications of conclusions that would be enthusiastically welcomed by the professional world as contributions to a knowledge of our economic and social environment." This material is not released, however, either because of ignorance of its general importance or because of fear of competitors. In addition to all this social research carried on by individual business firms, considerable research is also conducted by business trade associations, by commercial consultants and research organizations, and also by the Government and universities for business. Here again, in applied social science, we find the same collaboration among different kinds of organizations that occurs in natural science.

On the other side of the fence in business, that is, in the trade unions, a certain amount of social research is carried on also. But trade union leaders have, on the whole, been no more receptive to the use of social science than entrepreneurs and managers. In recent years the use of applied social science by trade unions has been encouraged "by the increased contacts of trade unions with Government; the need for frequent and well developed written presentations to various public and private agencies; the demands for a

based on sound conceptual schemes. Just the same situation
; for social science. However much its progress may be guided
ır social purposes, so long as the social science product is em-
llly tested and theoretically based, then we have genuine science.
ocial science, like all natural science, arises in part out of "prac-
social interests. Freudian psychology comes out of medical
py; some anthropology comes out of colonial administration;
sociology has come out of social reform movements; and some
cal science has come out of practical politics. Yet they all have
ntific validity apart from their origins. This is not to say that
factors cannot interfere with the development of social science;
s only to say that it is wrong to think that social science is *nec-*
ly "un-scientific" because of the social influences on it.
ne social factor which we have seen is always in interaction
science is social values. As with natural science, so also with
science, the two elements are often treated as if they were in
ımental opposition, or as if there were social science on the
and, divorced from values, and social values on the other hand.
s we have seen in the last chapter, we do not have so much a
ct between science and all social values as an interaction, and
le conflict, between different sets of values. As Professor Alex-
Leighton, the psychiatrist-anthropologist who was in charge
portant social research on Japanese morale during the recent
as said, "Social science does not 'threaten basic human values';
merely one among many forces threatening some values and
dogmas through upsetting the assumptions that underlie them.
values are strengthened."[28]
our society social science shares to some extent the generalized
approval that we give to all scientific and rational activities.
s a source of strength in its relations with other social interests
her social values. But this generalized approval probably does
tend in such great force to social science as it does to natural
e, and therefore social science is subject to much greater at-
When social science rationally investigates other social values,
hen it thereby seems to be undermining them because it is
teristic of "sacred" values in all spheres of social life not to
erly tolerant of "profane" rational examination, then social
falls under the danger of being restricted or even done away
ltogether. To put it concretely, Congressmen who do not want

critical evaluation of economic and social problems arising in negotiations; and wide recognition of the value of public support obtained through careful presentation of the union's case."[23] To some extent, unfortunately, this use of social scientists by labor unions to "prove a case" has led to strengthening of the sentiment that social science cannot really ever be objective.

Such problems are not, of course, peculiar to trade unions. The role of the applied social scientist presents certain typical difficulties in all organizations—Government, business, or trade union—which wish to employ social research in the formulation of social policy or in the administration of established social programs. These typical difficulties have been receiving more and more attention from the social scientists themselves.[24] While no completely satisfactory solutions are yet available, a certain amount of careful analysis of these typical difficulties has been made and its value is worth considering at this point.

The fundamental cause of the difficulty is the high degree of indeterminacy in most of social science knowledge. The social scientist acting as advisor to the policy-maker or the administrator cannot often present him with very reliable knowledge about the alternatives he faces. Hence, suggests Merton, the "ambivalence of distrust and hopeful expectation directed toward the social scientist in his capacity as advisor."[25] Hence also the possibility and even the likelihood that policy-maker and administrator will "distort" the social science research done for them in order to use it for their own pressing purposes. And hence, still further, the reluctance of various publics to support policies which are presumably relatively solidly supported by social science research. Since this relative indeterminacy of social science can only be eliminated very gradually, with the progress of basic social science, the difficulties we speak of can for the present only be minimized, not eradicated. They can be reduced in importance, that is, by proper caution on the part of social scientists to state the limitations of their work and to attempt some measure of control over the use of such work. Increasingly, for example, it may be desirable that social scientists should make their professional associations responsible for the formulation and maintenance of certain minimum standards of performance and "ethical" conduct. Sociologists and social psychologists specializing in the measurement and interpretation of public opinion have recently made an effort toward this kind of professional self-control.[26] Clinical psychologists, to take another example, have also been feeling more concern lately for professional discipline and self-control.[27] As social science becomes more effective, and therefore as its applications become "affected with the public interest," certain standards will have to be set up and maintained for such applications. Since these standards are in large part technical ones which can only be judged by the professionally competent, a close collaboration between public authority and the professional social science organizations is the most desirable way of achieving satisfactory public control. This is the procedure, of course, by which the medical profession maintains its standards of technical competence and ethical integrity.

There are still other difficulties in the way of integrating social science research into social decision besides those arising from its relative indeterminacy. For instance, the social scientist often suffers from confusion about his proper function in the organization which employs him, a confusion which is sometimes self-induced but also sometimes created by his administrative superiors. The social scientist who conceives of himself as more than a technician, who wishes to influence the policy which is to be finally chosen rather than merely "implementing" it, as the cant phrase goes, may find himself in conflict with the executive who does not wish to delegate or share any part of his prerogative and responsibility for making decisions. A preliminary requirement of all applied social science research, then, is that the social scientist search out the precise nature of his own values and functions and those of his "client" as well. From such understanding, which he should share with the client, he can minimize if not eliminate potential conflicts between himself and his administrator-client or administrator-superior. He can, for example, eliminate difficulties which are sometimes attributed wholly to failures of communication between himself and others. He can be more sure of what kind of information the administrator actually wants, that is, whether relatively "pure" research requiring considerable time is in order or whether only already available data must be quickly presented. Knowing his own function better, the social scientist can understand and comply with the time limitations and the other limitations which are always imposed on the administrator

either by his own superiors or by the events for w
decisions. The social scientist can, in short, when
staff capacity for an organization, understand
function for the organization are. Without und
cepting his role, he must inevitably cause difficul
the people he serves. If his values are such as to
unsatisfactory to him, he must in all good sens
and seek, either elsewhere in the organization or
role in which he can more fully realize his valu
scientist who wishes to influence policy in an o
cludes him from that function may find it easier
pose indirectly, by carrying on research elsewher
sity, research which may have to be taken into a
make policy in the organization he has left. Of
he does elsewhere may still not be recognized or
he wishes to turn into a certain course of action
case of the social scientist who wishes to influe
scientist who wishes simply to deliver technical
organizational roles which demand decisions a
decisions. Technical information and decision
some extent, of course, but since different socia
emphases on decision, the social scientist must
alternative roles.

We may turn now to some considerations
invention and discovery in social science. Most
about natural science in this regard in Chapter
science. There are certain differences, however
science, for instance, because of the relative la
conceptual schemes at present, the influence
edge on the course of progress is not so gr
science. External social factors are somewhat r
is, indeed, one of the common charges agai
that they are "too much" influenced by otl
those of already established knowledge. But
ceives the nature of science. The course of dev
we have seen, is partly influenced by social v
side science itself. This influence does not m
"scientific," so long as what is discovered is e

"short-haired women and long-haired men" looking into the "sacredness" of private family life will vote against the National Science Foundation if it includes the social sciences.

Opposition to social science and distress about its presumptions has not, of course, been limited to politicians and the man in the street. There are a goodly number of scholars in the arts and the humanities who feel that social science is particularly subversive of the values they hold. They feel that social science minimizes and even denies the reality and importance of the emotional, the moral, the artistic, and the esthetic aspects of human life. They feel that social science wishes to substitute for the *appreciation* of moral and esthetic values a fatal *analysis* which will murder the values it vivisects. In some cases these men have been driven by their horror of the supposedly annihilating attack of social science on cherished values into a counter-attack on the principle of rationality itself, into a denial of the possibility of any rational understanding of social life. This last-ditch defense, however extremist, has not been without a certain justification. For it has been characteristic of some elements in social science to overlook or deny the significance of the humanistic disciplines. Social science has deep roots in the positivistic bias of a great deal of nineteenth century social theory, and even now it has not entirely cast off the influence of this misconceived and limited understanding of human behavior.[29] There are still positivistic social scientists who ignore or deny the whole area of the moral-esthetic-emotional and who try to understand human behavior entirely in terms of man's rational orientation to the world. Not all that is not rational in human life is ignorance and error and irrationality; not all that is non-empirical is "unreal."

On this understanding, there is clearly no inevitable conflict between the social sciences and the humanities. The social sciences, like all science, are primarily concerned for analysis, prediction, and control of behavior and values; the humanities, for their synthesis and appreciation. Each of these is a necessary function in man's adjustment to his social existence. Neither approach to life can replace the other entirely. Therefore social scientists and scholars in the humanities can abandon such recrimination and conflict as exists between them to get on with the job of defining their complementary spheres of interest and competence. Each partly must go its own legitimate way. Each can also be of use to the other—the social sciences by developing systematic and valid new understandings of human behavior, the humanities by providing those insights which sometimes anticipate the future progress of the social sciences.

As for the defensiveness of social scientists, which is perhaps greater than that of other scholars, they will be bad social scientists if they do not see that social science must inevitably undergo resistance and even attack from other social activities and other social values. We have seen that this happens even to natural science. How much the greater will resistance be to social science, which is so much less strongly approved as yet and which often works its corrosive effects so directly. Indeed, this is almost certainly one of the most important reasons for the relatively retarded state of the social sciences, that their activities have not been approved because they throw so many social values into question. Yet social scientists will also be bad scientists if they do not point out how much in accord with our deepest values social science is. It is not true that social science has no values and is based on a fundamental moral relativism. Social science teaches moral relativism only in a limited sense, because it teaches it in the interests of the moral value of "critical rationality." There can be no relativism on that score; social science is necessarily absolute, or as absolute as values ever get, when it takes a stand on that moral value. There is, to be sure, a mistaken notion even among some social scientists themselves that they hold no values, that every human social pattern and every human activity is as good as any other, whether in other societies or in our own. But this fallacy can only be held among those social scientists who do not recognize the fundamental inter-relations of science, social and natural, and the basic values of our society. For all the disturbance social science now creates, and for all that it might cause—although this actual and potential harm may actually be less than is caused by our social ignorance and social incompetence—we must support the development of social science. It is a fundamental expression of our values and a fundamental necessity for their realization.

Science, we have said several times now, makes possible prediction and control. If we have more social science, then we shall have the possibility of more social control. How much social control can we have? How much shall we want to have? How much will our values allow? These are questions that must be answered in some fashion as much by those who are favorable to social science as by those who are hostile. Now whatever one's sentiments in this matter may be, there are two aspects to the problem of the control that science makes possible. There is, first, the question of whether partial understanding and partial control of social behavior is valuable at all, whether anything less than complete control of the whole of society will do us any good. And there is, second, the fear that many people have that partial control will actually lead us into complete control, and that social science therefore leads us down a primrose path to our destruction in a beehive society. Each of these two questions we may consider in its own right, although they are related as well.

There are some who feel that social science is of very small value because it can never give complete control over our social behavior. This view is taken on the assumption that the uncontrolled portions of our behavior will inevitably disrupt the areas in which we do have understanding and control. Yet this assumption does not hold for natural science and for the physical and biological phenomena it studies and controls. All science seeks to discover the determinate relations that hold between specified and partial aspects of the totality of the empirical world. Natural science has not yet and never will give us complete control of the physical and biological universe in which we live. Yet it is none the less immensely useful. To assume that knowledge must be complete to be effective is to misconceive the nature of rational knowledge. Human knowledge, whether consisting merely of rational "common sense" or of rational science, is always only one among the several things that influence social behavior. It is for that reason by its very nature limited.

On the same understanding, this is why we do not have to fear, either, that social science will create the possibility of complete social control. The satirists of the present totalitarian society, men like George Orwell, who assume that the totalitarian society has been completely controlled, or can ever be so controlled, are excessively credulous of the power of rational knowledge. Human society is inherently dynamic: a complex interaction among our values and our knowledge and our social organization and the physical and biological setting in which society is placed. A human beehive is impossible by the very nature of human society, by the very existence, for example, of changing social values which interact with and partially determine the uses of such rational knowledge as is available to men.

Of course, unhappily, there can be types of partial control which we would hate. There can be forms of partial control that may get us closer by far to the beehive society than we like. There can be the abhorrent "totalitarianism" of Nazi Germany and Communist Russia. But partial control just as much makes possible social conditions that are in accord with our values. Social science as much makes possible more "freedom" as more hateful "control." Knowledge is power to do good and evil alike, but we cannot throw away the power. We have faced the same dilemma in the consequences of the natural sciences, and we have chosen as we have had to choose, for the partial control that it gives us. We can give up our fear of a beehive society run by social science experts; and, having done so, we can devote ourselves to advancing social science to the point where it helps us bring our society a little more closely in accord with our social values. We have used natural science to give us relative abundance; social science can give us freedom in the same relative measure.

Whitehead has suggested that it was not until the seventeenth century that the conception of an Order of Nature became widespread among men in western society and therefore had a fructifying influence on the development of natural science. We may ask whether it is not possible that the twentieth century may mark the emergence of an analogous conception, the conception of an Order of Human Nature. If it does, a major advance in the development of human life will have occurred. If such a conception can be widely diffused through our society, partly by the gradual demonstration that social science is better than "common sense" for the prediction and control of human affairs, it will have a tremendously beneficial effect on the subsequent development of that very social science. It seems not at all unlikely that in the future we may witness a reciprocal process of social change in which social science earns support for itself by its achievements and in which these achievements in turn create a solid conviction of the Order of Human Nature. It seems not unlikely that we may gradually learn that human nature is no more arbitrary, capricious, chance-y, indeterminate, random, or inexplicable than physical nature or biological nature. Science in our society will not really achieve full maturity until social science comes of age with its sisters, the natural sciences.

Notes

CHAPTER I

1. P. W. Bridgman, "How far can scientific method determine the ends for which scientific discoveries are used?" *Social Science,* 22 (1947), p. 206. See also, P. W. Bridgman, *Reflections of a Physicist,* New York: Philosophical Library, 1950.

2. Anatol Rapoport, *Science and the Goals of Man,* New York: Harper & Bros., 1950.

3. Talcott Parsons, *The Structure of Social Action,* New York: McGraw-Hill Book Co., 1937.

4. J. B. Conant, *On Understanding Science,* New Haven: Yale University Press, 1947, and J. B. Conant, *Science and Common Sense,* New Haven: Yale University Press, 1951.

5. J. B. Conant, *On Understanding Science,* has noted this paradox, as has also Herbert Butterfield, *The Origins of Modern Science 1300-1800,* New York: The Macmillan Co., 1950, whose account I am more closely following here.

6. H. Butterfield, *The Origins.*

7. H. M. Johnson, "Pseudo-mathematics in the mental and social sciences," *American Journal of Psychology,* XLVIII (1936), 342-351.

8. W. B. Cannon, *The Wisdom of the Body,* New York: W. W. Norton, 1932.

9. Ellice McDonald, *Research and Its Organization,* Newark, Delaware: Biochemical Research Foundation, n.d. (c. 1950).

10. Anthony Standen, *Science is a Sacred Cow,* New York: E. F. Dutton and Co., 1950.

11. W. B. Cannon, *The Way of an Investigator,* New York: W. W. Norton, 1945, p. 35.

12. Cited in *ibid.,* p. 36.

13. H. Levy, *The Universe of Science,* New York: The Century Co., 1933.

14. J. B. Conant, *On Understanding Science,* p. 56.

15. René J. Dubos, *Louis Pasteur: Free Lance of Science,* Boston: Little, Brown and Co., 1950, p. 131.

16. Here again I am following H. Butterfield, *The Origins,* Ch. 5.

17. *Ibid.,* p. 79.

18. *Ibid.*

19. J. B. Conant, *On Understanding Science* and *Science and Common Sense.*

20. Thorstein Veblen, "The Evolution of the Scientific Point of View," in, *The Place of Science in Modern Civilization and Other Essays.* New York: The Viking Press, 1919.

21. Bergen Evans, *The Natural History of Nonsense,* New York: A. A. Knopf, 1946.

22. S. Lilley, "The commonsense of relativity," *Discovery,* March, 1949, p. 74.

23. *Ibid.*

24. Alexander Koyré, "The significance of the Newtonian synthesis," *Archives Internationales d'Histoire des Sciences,* 29 (1950), p. 304.

25. S. Lilley, "The commonsense," p. 74ff.

26. George Gamow, *Mr. Tompkins in Wonderland,* New York: The Macmillan Co., 1940. On the relation between modern physics and common sense, see also, P. W. Bridgman, "The scientist's commitment," *Bulletin of the Atomic Scientists,* 5 (1949), nos 6-7; and, Philipp Frank, "The place of logic and metaphysics in the advancement of modern science," *Philosophy of Science,* 15(1948), p. 277.

CHAPTER II

1. There are several one-volume histories of science available. See, for example, Sir Lawrence Bragg, *et al., The History of Science,* Glencoe, Ill.: The Free Press, 1951; H. T. Pledge, *Science Since 1500,* London: H. M. Stationery Office, 1939; F. Sherwood Taylor, *A Short History of Science and Scientific Thought,* New York: W. W. Norton & Co., 1949. See also, George Sarton, *The History of Science and the New Humanism,* Cambridge: Harvard University Press, 1937, and George Sarton, *The Life of Science,* New York: Henry Schuman, Inc., 1948.

2. Cited in James R. Newman's review of H. Butterfield, *The Origins,* in *Scientific American,* July, 1950.

3. This is Newman's phrase in the above review for the picture that is drawn in Butterfield's book, a book which he praises highly.

4. For a summary and critical review of the sociology of knowledge, see R. K. Merton, "The sociology of knowledge," Ch. VIII in his *Social Theory and Social Structure,* Glencoe, Ill.; The Free Press, 1949.

5. Benjamin Farrington, *Greek Science, Its Meaning for Us: I. Thales to Aristotle;* and B. Farrington, *Greek Science, Its Meaning For Us: II. Theophrastus to Galen,* London: Pelican Books, 1944, 1949; B. Hessen, "The Social and Economic Roots of Newton's 'Principia'," in *Science at the Cross Roads,* London: Kniga Ltd., 1931; James Gerald Crowther, *British Scientists of the Nineteenth Century,* London: K. Paul, Trench, Trubner & Co., 1935; J. G. Crowther, *Famous American Men of Science,* New York: W. W. Norton and Co., 1937; J. G. Crowther, *The Social Relations of Science,* New York: The Macmillan Co., 1941; J. D. Bernal, *The Social Functions of Science,* New York: The Macmillan Co., 1939; J. D. Bernal, *The Freedom of Necessity,* London: Routledge and Kegan Paul, 1949; Lancelot Hogben, *Science for the Citizen,* London: G. Allen & Unwin, 1938; L. Hogben, *Mathematics for the Million,* New York: W. W. Norton and Co., 1940, rev. ed.; Dirk J. Struik, *Yankee Science in the Making,* Boston: Little, Brown and Co., 1948; W. Salant, "Science and society in ancient Rome," *Scientific Monthly,* 47(1938), 525-35.

6. I have here been following S. Lilley, *Men, Machines and History,* London: Cobbett Press, 1948, pp. 417ff.

7. S. Lilley, "Social aspects of the history of science," *Archives Internationales d'Histoire des Sciences,* 28 (1948-49), p. 385.

8. *Ibid.*

9. J. G. Crowther, *The Social Relations of Science,* p. 10.

10. S. Lilley, *Men, Machines,* Ch. I. In my account of early science I am most directly indebted to Lilley and to: R. J. Forbes, *Man the Maker, A History of Technology and Engineering,* New York: Henry Schuman, 1950; R. J. Forbes, "Man and matter in the ancient Near East," *Archives Internationales d'Histoire des Sciences,* 27(1947-48), 557-573; and R. J. Forbes, "The ancients and the machine," *Archives Internationales d'Histoire des Sciences,* 28(1948-49), 919-33.

11. S. Lilley, *Men, Machines,* Ch. I.

12. Dirk J. Struik, "Stone age mathematics," *Scientific American,* 179(1948), Dec., p. 44.

13. See esp. Bronislaw Malinowski, *Magic, Science, and Religion,* Glencoe, Ill.: The Free Press, 1948, Ch. I.

14. O. T. Mason, *The Origins of Invention: A Study of Industry Among Primitive Peoples,* New York: Charles Scribner's Sons, 1915, p. 410.

15. M. F. Ashley Montagu, "Primitive medicine," *The Technology Review,* Nov., 1945, 27-30, 56.

16. A. L. Kroeber, "The Eskimos as aboriginal inventors," *Scientific American,* 110(1914), p. 54.

17. B. Malinowski, *Magic, Science,* p. 17.

18. *Ibid.*

19. My analysis of magic and science rests on B. Malinowski, *Magic, Science* and on unpublished lectures by Prof. Talcott Parsons.

20. F. S. Taylor, *A Short History,* p. 21.

21. The facts, but not the interpretation, of the following account are drawn chiefly from B. Farrington, *Greek Science,* I. and II., and R. J. Forbes, "The ancients and the machine."

22. B. Farrington, *Greek Science,* II., p. 10.

23. S. Lilley, *Men, Machines,* Ch. III.

24. B. Farrington, *Greek Science,* II., p. 15.

25. *Ibid.,* p. 28.

26. On Aristotle's experimentation, see Richard P. McKeon, "Aristotle and the origins of science in the West," in R. C. Stauffer, ed., *Science and Civilization,* Madison: University of Wisconsin Press, 1949, pp. 13-14. See also, Otto Blüh, "Did the Greeks perform experiments?" *American Journal of Physics,* 17(1949), pp. 384-388.

27. B. Farrington, *Greek Science,* II., Ch. II.

28. Gilbert Murray, *Five Stages of Greek Religion,* London: Watts & Co., 1946, p. 128. See also, Carl B. Boyer, "Aristotle's Physics," *Scientific American,* May, 1950, 48-51, for an emphasis on the "inner consistence" principle.

29. B. Farrington, *Greek Science,* II., p. 164.

30. See R. J. Forbes, "Professions and crafts in Ancient Egypt," *Archives Internationales d'Histoire des Sciences,* 3(1950), 599-618, for evidence that the class structure of ancient Egypt and Mesopotamia was somewhat less simple, somewhat less a "slave society" than is often alleged. "This," he says, "should make us wary when blaming social conditions for certain aspects of scientific and technological evolution." p. 618.

31. The following discussion is a paraphrase and quotation from A. N. Whitehead, *Science and the Modern World,* New York: The Macmillan Co., 1925, pp. 17-19.

32. On Christian naturalism in the Mediaeval Period, see William M. Agar, *Catholicism and the Progress of Science,* New York: The Macmillan Co., 1940, p. 14. See also, Lynn White, Jr., "Technology and invention in the Middle Ages," *Speculum, a Journal of Mediaeval Studies,* 15(1940), 141-159.

33. S. Lilley, *Men, Machines,* p. 190.

34. Martha Ornstein, *The Role of the Scientific Societies in the Seventeenth Century,* Chicago: University of Chicago Press, 1938, p. 50, speaks of a "mutation."

35. *The Origins,* p. 174.

36. *The Origins,* p. 172. See also, G. N. Clark, *Science and Social Welfare in the Age of Newton,* rev. ed., Oxford: The Clarendon Press, 1949, p. 74, for the same view.

37. *The Origins,* p. 169.

38. Morris R. Cohen, "Descartes," *Encyclopedia of the Social Sciences.*

39. *The Origins,* p. 77.

40. A. N. Whitehead, *Science,* p. 3.

41. In what follows I am greatly indebted to Martha Ornstein's classic book, *The Role of the Scientific Societies.*

42. Harcourt Brown, *Scientific Organizations in Seventeenth Century France (1620-1680),* Baltimore: The Williams and Wilkins Co., 1934, *pas-*

sim, but esp. at p. 56, gives many details of the visits and letters among the French, Italian, British, and Dutch scientists of the seventeenth century.

43. G. N. Clark, *Science and Social Welfare,* p. 24.

44. S. Lilley, "Social aspects," pp. 386ff.

45. *Ibid.*

46. Max Weber, *The Protestant Ethic and the Spirit of Capitalism,* trans. by T. Parsons, London: George Allen & Unwin, 1930. R. K. Merton, *Science, Technology and Society in Seventeenth Century England, OSIRIS,* IV, part 2, Bruges (Belgium), 1938.

47. Max Weber, *The Religion of China,* trans. by H. H. Gerth, Glencoe, Ill.: The Free Press, 1951, Ch. VI and pp. 196ff.

48. See Jean Pelseneer, "L'origine Protestante de la science moderne," *LYCHNOS,* 1946-47, 246-248, for a statistical study showing that the number of scientists in the sixteenth century who are attached to Protestantism is greater than those remaining faithful to Catholicism. Says Pelseneer, "Nous concluerons donc que la science moderne est née de la Reforme."

49. R. K. Merton, "Puritanism, Pietism and Science," Ch. XIV in his *Social Theory and Social Structure.*

CHAPTER III

1. Talcott Parsons, *Social Science: A Basic National Resource,* Unpublished manuscript, 1949. Also, Talcott Parsons, *The Social System,* Glencoe, Ill.: The Free Press, 1951, Ch. VIII.

2. An excellent analysis of the origins and nature of the values of liberal democratic society may be found in A. D. Lindsay, *The Modern Democratic State,* New York: The Oxford University Press, 1947.

3. W. E. Moore, *Industrial Relations and the Social Order,* New York: The Macmillan Co., 1946, gives these facts on pp. 57-58.

4. See *Benjamin Franklin's Experiments: a new edition of Franklin's "Experiments and observations on electricity,"* edited with a critical and historical introduction by I. Bernard Cohen, Cambridge: Harvard University Press, 1941.

5. S. Giedion, *Mechanization Takes Command,* New York: Oxford University Press, 1948.

6. The erroneous view that science can flourish equally well under *any* type of government has been taken by George Lundberg, *Can Science Save Us?,* New York: Longmans, Green and Co., 1947. See also, Frank H. Knight, "Salvation by science: the gospel according to Professor Lundberg," *Journal of Political Economy,* 55 (1947), 537-52.

7. J. B. Conant, "Science and politics in the twentieth century," *Foreign Affairs,* 28 (1950), p. 189.

8. See Leslie E. Simon, *German Research in World War II. An Analysis of the Conduct of Research,* New York: John Wiley & Sons, 1947; and, Samuel A. Goudsmit, *Alsos,* New York: Henry Schuman, 1947.

9. Joseph Needham, *The Nazi Attack on International Science,* London: Watts and Cox, 1941.

10. J. Needham, *The Nazi Attack, passim.* See also, E. Y. Hartshorne, *German Universities and National Socialism,* Cambridge: Harvard University Press, 1937; and Robert K. Merton, "Science and the Social Order," Ch. XI in his *Social Theory and Social Structure.*

11. Sir Richard Gregory, *Science in Chains,* London: Macmillan Co., 1941, quotes this statement.

12. Sir R. Gregory, *Science in Chains.*

13. See L. E. Simon, *German Research;* S. A. Goudsmit, *Alsos;* and James Phinney Baxter, *Scientists Against Time,* Boston: Little, Brown and Co., 1946, pp. 7ff.

14. L. E. Simon, *German Research,* p. 48.

15. *Ibid.*

16. Eric Ashby, *Scientist in Russia,* New York: Penguin Books, 1947, esp. pp. 17-29, 83, 99. Ashby was in Russia during the war and found much to admire in Soviet science and much that was identical, in its organization and day-to-day activity, to British science. See also, H. J. Muller, "The destruction of science in the USSR" and "Back to barbarism—scientifically," *The Saturday Review of Literature,* Dec. 4, Dec. 11, 1948. Muller, who had himself worked in Russia, reports that Soviet genetics had reached a high state of advancement by 1935, after which it was undermined by Lysenkoism.

17. For a description of the Soviet conception of the essential rationality of human nature and for an excellent analysis of the functions of this conception in Soviet society, see Raymond A. Bauer, *The New Man in Soviet Psychology,* Cambridge: Harvard University Press, 1952.

18. J. B. Conant, "Science and politics," p. 190.

19. W. W. Leontieff, Sr., "Scientific and technological research in Russia," *American Slavic and East European Review,* 4(1945), p. 70. No comprehensive figures are available for the years after 1935, but Leontieff thinks "there is every good reason to believe that the 1929-1935 rate of expansion was maintained and even increased in the later years."

20. W. W. Leontieff, Sr., "Scientific and technological research," p. 70. Further on the organization of Soviet science, see Eric Ashby, *Scientist in Russia;* Irving Langmuir, "Science and incentives in Russia," *The Scientific Monthly,* 63(1946), 85-92; Gerald Oster, "Scientific research in the USSR. Organization and Planning," *Annals* of the A.A.P.S.S., 263(1949), 134-139; and, Nancy Mazur, *Science in Soviet Russia: A Sociological Analysis,* Undergraduate Honors Thesis, Smith College, 1951.

21. Alex Inkeles, "Stratification and mobility in the Soviet Union," *American Sociological Review,* 15 (1950), 465-479.

22. J. B. Conant, "Science and politics."

23. Josef Brozek, "Extension of political domination beyond Soviet genetics," *Science,* 111(1950), 389-91.

24. There has been a plethora of writings about Lysenkoism in Soviet genetics. Much the best complete discussion may be found in Julian Huxley, *Heredity East and West; Lysenko and World Science,* New York: Henry Schuman Co., 1949. See esp. his summary of *factors* accounting for Lysenkoism at p. 187. See also, Bertram D. Wolfe, "Science joins the party," *Antioch Review,* X(1950), 47-60; Conway Zirkle, ed., *Death of a Science in Russia,* Philadelphia: Univ. of Penna. Press, 1949; and Eric Ashby, *Scientist in Russia,* pp. 114-117.

25. For a keen awareness of the necessity for "pure" science, see the excerpt from the article by the famous physicist, Peter Kapitsa, member of the Russian Academy of Sciences, quoted in W. W. Leontieff, Sr., "Scientific and technological research," pp. 73ff., esp. p. 75.

CHAPTER IV

1. This basic point about the nature of science has been insisted upon very strongly by P. W. Bridgman, "The scientist's commitment." See also, Michael Polanyi, "Freedom in science," *Bulletin of the Atomic Scientists,* VI (1950), 195-198, 224; Michael Polanyi, *Science, Faith, and Society,* New York: Oxford University Press, 1948; and, Max Weber, "Science as a Vocation," in *From Max Weber: Essays in Sociology,* ed. and trans. by H. H.

Gerth and C. W. Mills, New York: Oxford University Press, 1946.

2. P. W. Bridgman, *Reflections of A Physicist*, p. 318.

3. See George Sarton, *The History of Science*, pp. 14, 29, for an expression of this aspect of scientific faith.

4. Claude Bernard, *An Introduction to the Study of Experimental Medicine*, trans. by Henry C. Greene, New York: Henry Schuman, 1949, p. 165.

5. René Dubos, *Louis Pasteur*, p. 72.

6. Leopold Infeld, *Quest: The Evolution of a Scientist*, New York: Doubleday, Doran, 1941, p. 280.

7. This is what R. K. Merton calls the value of "communism," a term which seems less desirable today than when he first used it because of its political and ideological significance. See his "Science and Democratic Social Structure," Ch. XII in his *Social Theory and Social Structure*.

8. For an argument against secrecy in terms of its inexpediency, see Louis N. Ridenour, "Secrecy in science," *Bulletin of the Atomic Scientists*, 1 (1946), no. 6; and Walter Gellhorn, *Security, Loyalty, and Science*, Ithaca, N. Y.: Cornell University Press, 1950.

9. *Fortune Magazine*, "The scientists," October, 1948, 106-112, 166-76.

10. Talcott Parsons, "The Professions and Social Structure," Ch. VIII in his *Essays in Sociological Theory*, Glencoe, Ill.: The Free Press, 1949. See also Talcott Parsons, *The Social System*.

11. Archie M. Palmer, *Survey of University Patent Policies*, Washington, D.C.: National Research Council, 1948.

12. Edmund W. Sinnott, "Ten million scientists," *Science*, 111 (1950), 123-129.

13. C. E. K. Mees, *The Path of Science*, New York: John Wiley and Sons, 1946. John R. Baker, *Science and the Planned State*, New York: The Macmillan Co., 1945, p. 17.

14. U. S. National Resources Planning Board, *Research—A National Resource: II. Industrial Research*, Washington, D.C.: U. S. Government Printing Office, 1940, p. 99.

15. *Fortune Magazine*, "Nylon," 22, July, 1940, 57-60, 114-116.

16. Gerard Piel, "Mathematics comes out of the classroom," *Yale Review*, XXXIX (1949), Autumn, 132-141.

17. For an extremely naive glorification of the motives of "pure" scientists, see Sir Richard Gregory, *Discovery, Or The Spirit and Service of Science*, New York: The Macmillan Co., 1929, pp. 6, 9, 11, 12, 27, 50, 55. For a more realistic statement, see G. H. Hardy, *A Mathematician's Apology*, Cambridge: Cambridge University Press, 1941, p. 18.

18. In E. P. Wigner, ed., *Physical Science and Human Values*, Princeton: Princeton University Press, 1947, p. 142.

19. Letter, *Science*, 100 (1944), p. 471.

20. J. B. Conant, *On Understanding Science*, pp. 22-23. Also, J. B. Conant, "The scientist in our unique society," *The Atlantic Monthly*, March, 1948, 47-51.

CHAPTER V

1. U. S. President's Scientific Research Board, *A Program for the Nation*, Vol. One of *Science and Public Policy*, Washington, D.C.: U. S. Government Printing Office, August 27, 1947, p. 9.

2. C. C. North and Paul K. Hatt, "Jobs and occupations: a popular evaluation," in L. Wilson and W. L. Kolb, *Sociological Analysis*, New York: Harcourt, Brace, 1949.

3. See Robert F. Griggs, "Shall biologists set up a national institute?" *Science*, 105 (1947), 559-565.

4. *Science*, 97 (1943), p. 63.

5. *Fortune Magazine*, "The scientists."

6. *Science*, 101 (1945), p. 240.

7. U. S. President's Scientific Research Board, *Administration for Research*, Vol. Three of *Science and Public Pol-*

icy, Washington, D.C.: U. S. Government Printing Office, Oct. 4, 1947, p. 206.

8. For an unfavorable comment on one system of distinctions in science, the system of starring "outstanding" men in the who's who of science, *American Men of Science,* see *Science,* 101(1945), p. 639. For a description of the procedure employed in this system, see S. S. Visher, *Scientists Starred, 1903-1943, in American Men of Science,* Baltimore: The Johns Hopkins University Press, 1947; and, *Science,* 99(1944), 533-544.

9. George W. Gray, "The Nobel Prizes," *Scientific American,* 181, no. 6 (1949), 11-17.

10. P.S.R.B., *Administration for Research,* App. III.

11. *Ibid.,* p. 205.

12. *Ibid.,* p. 206.

13. *Ibid.,* p. 205.

14. *Ibid.,* p. 205.

15. *Ibid.,* p. 206.

16. *Ibid.,* p. 223.

17. *Fortune Magazine,* "The scientists," p. 174.

18. Eric Ashby, *Scientist in Russia,* Ch. 8, esp. at p. 192.

19. Francis Hughes, "Soviet invention awards," *Economic Journal,* 55 (1945), 291-97; Francis Hughes, "Incentive for Soviet initiative," *Economic Journal,* 56(1946), 415-425; Francis Hughes, "Incentives and the Soviet inventor," *Discovery,* January, 1947, 10-12, 32; and Ervin O. Anderson, "Nationalization and international patent problems," *Law and Contemporary Problems,* XII(1947), 782-795.

20. M. Ornstein, *The Role of the Scientific Societies;* H. Brown, *Scientific Organizations;* and Dorothy Stimson, *Scientists and Amateurs: A History of the Royal Society,* New York: Henry Schuman, 1948.

21. Richard H. Shryock, "American indifference to basic science during the nineteenth century," *Archives Internationales d'Histoire des Sciences,* 28 (1948-49), 50-65; and Richard H. Shryock, "Trends in American medical research during the nineteenth century," *Proc. of the American Philosophical Society,* 19(1947), 58-63.

22. J. R. Baker, *Science and the Planned State,* p. 18.

23. D. J. Struik, *Yankee Science in the Making,* passim, esp. pp. 322, 341.

24. U. S. National Resources Committee, *Research—A National Resource. I. Relation of the Federal Government to Research,* Washington, D. C.: U. S. Government Printing Office, 1938, p. 168.

25. S. S. Visher, *Scientists Starred.*

26. Norbert Wiener, *Cybernetics or Control and Communication in the Animal and the Machine,* New York: John Wiley & Sons, 1948, p. 8.

27. British Association of Scientific Workers, *Science and the Nation,* London: Pelican Books, 1947, p. 193.

28. Cited in Ralph S. Bates, *Scientific Societies in the United States,* New York: John Wiley and Sons, 1945, p. 3.

29. For a list of these, see *Ibid.*

30. F. R. Moulton, "The AAAS and Organized American Science," *Science,* 108(1948), 573-77.

31. Clifford Grobstein, "The Federation of American Scientists," *Bulletin of the Atomic Scientists,* VI(1950), pp. 58, 61.

32. *Ibid.*

33. Bernard Barber, "Participation and 'Mass Apathy' in Associations," in A. W. Gouldner, ed., *Studies in Leadership,* New York: Harper & Bros., 1950.

34. W. Stephen Thomas, *The Amateur Scientist,* New York: W. W. Norton, 1942.

35. On the types of scientific research possible for amateurs, see also J. B. S. Haldane, *Possible Worlds. A Scientist Looks at Science,* New York: Harper & Bros., 1928, Ch. XXIV.

36. W. S. Thomas, *The Amateur Scientist,* p. 169.

37. For a very vivid concrete descrip-

tion of some aspects of this informal organization of science, with especial reference to the influence of key scientists on job assignments, see L. Infeld, *Quest,* pp. 294ff.

38. *Fortune Magazine,* "The great science debate," 33(1946), June, 116-123, 236ff. There was, of course, some opposition to this "scientific oligarchy". See Talcott Parsons, "The science legislation and the role of the social sciences," *American Sociological Review,* XI(1946), p. 657.

39. Vannevar Bush, *Modern Arms and Free Men,* New York: Simon and Schuster, 1949, p. 3.

40. Merriam H. Trytten, "The Mobilization of Scientists," in Leonard D. White, ed., *Civil Service in Wartime,* Chicago: The University of Chicago Press, 1945.

41. *Ibid.,* p. 60.

42. For other indications of the importance of personal contacts among men with previous acquaintance and mutual confidence, see J. P. Baxter, *Scientists Against Time,* pp. 14ff.

43. V. Bush, *Modern Arms,* p. 87.

44. *Fortune Magazine,* "Great science debate," p. 120. On this same point, see J. P. Baxter, *Scientists Against Time,* p. 18.

45. For a description of similar patterns of informal control in British science, see J. D. Bernal, *The Social Functions of Science,* Ch. 3.

46. C. E. Ayres, *Huxley,* New York: W. W. Norton & Co., 1932.

47. Wilson F. Harwood, "Budgeting and cost accounting in research institutions," in George P. Bush and Lowell H. Hattery, eds., *Scientific Research: Its Administration and Organization,* Washington, D. C.: The American University Press, 1950.

48. On the larger movement, on some of the social problems it has caused, and on a democratic program for dealing with these problems, see Karl Mannheim, *Freedom, Power & Democratic Planning,* New York: Oxford University Press, 1950.

49. See Robert K. Merton, *et al., Reader in Bureaucracy,* Glencoe, Ill.: The Free Press, 1952.

50. J. P. Baxter, *Scientists Against Time,* p. 20.

51. F. B. Jewett, in Standard Oil Development Co., *The Future of Industrial Research,* privately printed, 1945, p. 18.

52. E. McDonald, *Research and Its Organization.*

53. For this and other examples, see C. E. K. Mees, *The Path of Science,* pp. 177-178.

54. In E. P. Wigner, ed., *Physical Science,* p. 51.

55. *Ibid.,* p. 54.

56. L. Kowarski, "Psychology and structure of large-scale physical research," *Bulletin of the Atomic Scientists,* 5(1949), nos. 6-7, p. 187.

57. R. K. Merton, *Reader in Bureaucracy.*

58. C. E. K. Mees, in E. P. Wigner, ed., *Physical Science,* p. 59.

59. Frank B. Jewett, "The promise of technology," *Science,* 99(1944), p. 5.

60. On the functions of the director of a large scientific research organization, see also Ellice McDonald, *Research and Its Organization.*

61. Lowell H. Hattery, "New Challenge in Administration," in G. P. Bush and L. H. Hattery, eds., *Scientific Research.*

62. Philip N. Powers, "The changing manpower picture," *Scientific Monthly,* LXX(1950), 165-171.

63. P.S.R.B., *Administration for Research.*

64. P. N. Powers, "The changing manpower picture," p. 165.

65. *Ibid.*

66. *Ibid.,* p. 170.

67. *Science,* 24(1906), 732-44.

68. *Fortune Magazine,* "The scientists," pp. 110ff.

69. M. H. Trytten, "Significance of

Fellowships in Training Scientific Workers," in G. P. Bush and L. H. Hattery, eds., *Scientific Research.*

70. See William Miller's four articles: "American historians and the business elite," *The Journal of Economic History,* IX(1949), 184-208; "The recruitment of the American business elite," *Quarterly Journal of Economics,* LXIV(1950), 242-53; "American lawyers in business and politics. Their social backgrounds and early training," *The Yale Law Journal,* January, 1951, 66-76; and "The Business Elite in Business Bureaucracies," in William Miller, ed., *Men in Business,* Cambridge: Harvard University Press, 1952. See also, H. B. Goodrich, *et al.,* "The origins of U. S. scientists," *Scientific American,* July, 1951, 15-17.

71. J. M. Cooper, "Catholics and scientific research," *Commonweal,* 42 (1945), 147-49; Prof. James A. Reyniers, Director of the Lobund Laboratories at Notre Dame University, reported in the New York Times, Dec. 12, 1949; and Joseph P. Fitzpatrick, S.J., "Catholic responsibilities in sociology," *Thought, Fordham University Quarterly,* XXVI(1951), 384-96.

72. U. S. Department of Labor, Women's Bureau, *The Outlook for Women in Science,* Washington, D. C.: U. S. Government Printing Office, 1949.

73. John Mills, *The Engineer in Society,* New York: D. Van Nostrand Co., 1946, p. 113.

74. U. S. Dept. of Labor, Women's Bureau, *The Outlook for Women in Science,* p. 52.

75. Elizabeth Wagner Reed, "Productivity and attitudes of seventy scientific women," *American Scientist,* 38 (1950), 132-135.

CHAPTER VI

1. On the small amount of time left for research to the faculties of American colleges in the early nineteenth century, see R. S. Bates, *Scientific Societies,* p. 30.

2. Thorstein Veblen, *The Higher Learning in America,* New York: B. W. Huebsch, 1918, p. 15.

3. U. S. National Resources Committee, I. *Relation of the Federal Government to Research,* p. 170.

4. T. Veblen, *The Higher Learning,* p. 33.

5. W. B. Cannon, *The Way of an Investigator,* p. 79.

6. *Ibid.,* Ch. VI, "Passing on the Torch," for a description of this process.

7. Logan Wilson, *The Academic Man,* New York: Oxford University Press, 1942, p. 33.

8. U. S. National Resources Committee, I. *Relation of the Federal Government to Research,* Sec. 6.

9. Several studies have shown that only a small proportion of all Ph.D. degree holders, say, about 25%, continue their research activities after completing their thesis. See *Ibid.,* p. 171.

10. Vannevar Bush, *Science, The Endless Frontier,* Washington, D. C.: U. S. Government Printing Office, 1945. For further evidence of the importance of the small colleges, see H. B. Goodrich, *et al.,* "The origins of U. S. scientists."

11. Carl G. Hartman, "The little researcher," *Science,* 103(1946), 493-96.

12. John E. Flynn, "Research activity in the smaller institutions: An open letter from an editor," *American Scientist,* 30(1942), 303-305.

13. Talcott Parsons, unpublished lectures. See also James B. Conant, *The President's Report,* 1948, Harvard University, p. 1ff, for a similar statement.

14. Kingsley Davis, *Human Society,*

New York: The Macmillan Co., 1949, p. 445.

15. H. Levy, *The Universe of Science,* p. 197.

16. T. Veblen, *The Higher Learning,* p. 99.

17. On this divergence between ideal and reality, see L. Wilson, *The Academic Man.*

18. For a discussion of the problem of blending these two in university tenure appointments, see James B. Conant, "Academical Patronage and Superintendence," Occasional pamphlets of the Graduate School of Education, Harvard University, no. 3, 1938.

19. For a discussion of these problems in their general form, see Florian Znaniecki, *The Social Role of the Man of Knowledge,* New York: Columbia University Press, 1940; for a study of their specific forms in the U. S., see L. Wilson, *The Academic Man,* pp. 217ff.

20. P.S.R.B., *A Program for the Nation,* p. 4.

21. A. L. Kroeber, *Configurations of Culture Growth,* Berkeley: University of California Press, 1944, p. 166.

22. The list has been taken from *Fortune Magazine,* "The great science debate," p. 240.

23. A. L. Kroeber, *Configurations,* p. 166.

24. *Fortune Magazine,* "The great science debate," p. 240. On this point, see also the testimony by Professor Harold C. Vrey, Nobel Prize winner, before the Senate Subcommittee on War Mobilization in 1945, reported in James R. Newman and Byron S. Miller, *The Control of Atomic Energy,* New York: McGraw-Hill Book Co., 1948, p. 179.

25. R. H. Shryock, "Trends in American medical research" and "American indifference to basic science."

26. P.S.R.B., *Administration for Research,* Ch. VIII.

27. The following facts are from the *New York Times,* Dec. 5, 1949.

28. Archie M. Palmer, *Survey of University Patent Policies,* Ch. VI.

29. C. E. K. Mees, *The Path of Science,* p. 185.

30. A. M. Palmer, *Survey of University Patent Policies,* p. 8.

31. *Ibid.,* p. 150.

32. *Ibid.,* p. 151.

33. See the Research Corporation pamphlets, *A Review for the Period, 1912-1945,* and, *Grants-in-aid of Scientific Research and Fields of Patent Interest,* published by the Research Corporation, New York, 1945, 1947.

34. See E. V. Hollis, *Philanthropic Foundations and Higher Education,* New York: Columbia University Press, 1938, pp. 239ff. Also on the foundations, see F. P. Keppel, *The Foundation; its place in American life,* New York: The Macmillan Co., 1930; W. S. Rich and N. R. Deardorff, *American Foundations and Their Fields,* VI, New York: Raymond Rich Associates, 1948; E. C. Lindeman, *Wealth and Culture; a study of one hundred foundations,* New York: Harcourt, Brace, 1936; and, Shelby M. Harrison and F. Emerson Andrews, *American Foundations for Social Welfare,* New York: Russell Sage Foundation, 1946.

CHAPTER VII

1. Quoted in F. Russell Bichowsky, *Industrial Research,* Brooklyn, N. Y.: Chemical Publishing Co., 1942, p. 70. On the importance of these functions of the Research Director, see also C. E. K. Mees, *The Path of Science,* p. 218.

2. U. S. National Resources Planning Board, II. *Industrial Research,* p. 120.

3. See U. S. National Resources Planning Board, *Research—A National Resource.* III. *Business Research,* Washington D.C.: U. S. Government Printing Office, 1941, for a discussion of the industrial uses of social science.

4. Frank B. Jewett, "Research in industry," *Scientific Monthly,* 48(1939), 195-202.

5. U. S. National Resources Planning Board, II. *Industrial Research,* Sec. IV, p. 176.

6. *The Technology Review,* Feb., 1949, editorial notes. On the engineering profession in general, see Theodore J. Hoover and John C. L. Fish, *The Engineering Profession,* Palo Alto: Stanford University Press, 1947.

7. U. S. National Resources Planning Board, II. *Industrial Research,* p. 182.

8. *Ibid.,* p. 186.

9 C. E. K. Mees, *The Path of Science,* p. 198.

10. U. S. National Resources Planning Board, II. *Industrial Research,* pp. 88ff.

11. C. E. K. Mees, *The Path of Science,* p. 14.

12. This is the summary view of the U. S. National Resources Planning Board, II. *Industrial Research,* p. 86.

13. See David B. Hertz, *The Theory and Practice of Industrial Research,* New York: McGraw-Hill Book Co., 1950, Chs. 7, 11, for a good discussion of "costing" problems in industrial research.

14. T. A. Boyd, *Research, The pathfinder of science and industry,* New York: D. Appleton-Century Co., 1935, p. 178.

15. See D. H. Killeffer, *The Genius of Industrial Research,* New York: Reinhold Co., 1948, Ch. 11, for a discussion of "The Pilot Plant and What Happens There."

16. C. E. K. Mees, *The Path of Science,* p. 224.

17. Standard Oil of California, *The Coordination of Motive, Men and Money in Industrial Research,* privately printed, 1946, p. 62. See also D. H. Killeffer, *The Genius of Industrial Research,* Ch. 13, on the technical problems of evaluating the profitability of research in the chemical industry.

18. F. R. Bichowsky, *Industrial Research,* pp. 113ff.

19. Standard Oil of California, *The Coordination of Motive, Men and Money in Industrial Research,* p. 53.

20. Robert E. Wilson, "Incentives for research," *The Technology Review,* Feb., 1947, 217-291, 232.

21. F. R. Bichowsky, *Industrial Research,* p. 111. On the "freedom" he himself felt in this situation, see Irving Langmuir, "Atomic hydrogen as an aid to industrial research," *The Scientific Monthly,* LXX(1950), 3-9.

22. See F. B. Jewett, "Research in industry," p. 197, for an account of some difficulties in establishing the early industrial research organizations.

23. For an elaborate description of the formal organization and management practices of a "typical" large research organization, see Standard Oil of California, *The Coordination of Motive, Men and Money in Industrial Research.* This handbook on how to organize a working industrial research department has many blueprint organization charts and detailed management guides for each position in such an organization. For other charts of this kind, see C. E. K. Mees, *The Path of Science,* pp. 190-192.

24. W. Rupert Maclaurin, *Invention and Innovation in the Radio Industry,* New York: The Macmillan Co., 1949, p. 163.

25. *Ibid.*

26. For the longest, but not fully satisfactory study of the actual workings of an industrial research laboratory, the Bell Telephone Laboratories, see J. Mills, *The Engineer in Society.*

CHAPTER VIII

1. T. Swann Harding, *Two Blades of Grass, A History of Scientific Development in the U. S. Department of Agriculture,* Norman: University of

Oklahoma Press, 1947, p. 6.

2. P.S.R.B., *Administration for Research,* Ch. VII.

3. P.S.R.B., *A Program for the Nation,* p. 45.

4. T. S. Harding, *Two Blades of Grass,* Ch. VIII. This book is an impressive account of the vast scope and tremendous value of the Department of Agriculture's scientific work during the last two hundred years. For a very detailed statement of the kinds of research now being carried on by sixteen different Federal agencies, see U. S. President's Scientific Research Board, *The Federal Research Program,* Vol. Two of *Science and Public Policy,* Washington, D.C.: U. S. Government Printing Office, Sept. 27, 1947. See also U. S. Department of the Army, *Scientists in Uniform—World War II,* Washington, D.C.: U. S. Government Printing Office, 1948, for a description of the uses of scientists in the armed forces during World War II.

5. U. S. National Resources Planning Board, I. *Relation of the Federal Government to Research,* p. 9.

6. P.S.R.B., *A Program for the Nation,* pp. 45ff.

7. U. S. National Resources Planning Board, I. *Relation of the Federal Government to Research,* p. 7.

8. *Ibid.,* p. 44.

9. P.S.R.B., *Administration for Research,* p. 142. See also, U. S. President's Scientific Research Board, *Manpower for Research,* Vol. Four of *Science and Public Policy,* Washington, D.C.: U. S. Government Printing Office, Oct. 11, 1947.

10. U. S. President's Scientific Reserach Board, *The Nation's Medical Research,* Vol. Five of *Science and Public Policy,* Washington, D.C.: U. S. Government Printing Office, Oct. 18, 1947, *passim.*

11. P.S.R.B., *A Program for the Nation,* p. 53.

12. M. H. Trytten, "The Mobilization of Scientists," p. 68. See also, pp. 66-67.

13. Edward U. Condon, "Recruitment and Selection of the Research Worker," in G. P. Bush and L. H. Hattery, *Scientific Research.*

15. Philip N. Powers, "The science training group in the Washington area," *Science,* 104(1946), p. 477.

15. Rawson Bennett, "Educational programs at research centers," *Science,* 106(1947), 283-284.

16. *Ibid.* See also David B. Dill, "Biologists in military service," *Science,* 111(1950), 675-676, for an account of how the Medical Division, Army Chemical Corps., has organized a Government research agency which provides an excellent place for research biologists and medical scientists to work.

17. P.S.R.B., *Administration for Research,* p. 108.

18. W. Gellhorn, *Security, Loyalty and Science;* and, *Fortune Magazine,* "The scientists," pp. 166ff.

19. P.S.R.B., *Administration for Research,* p. 136.

20. On O.N.R. see A. H. Hausrath, "Programs for fuller utilization of present resources of scientific personnel," *Science,* 107(1948), 360-363. There are great differences of opinion among American scientists about the effects of military support on American science as a whole. For an expression of some of these differences, see Louis N. Ridenour, "Military support of American science, a danger?", reprinted from *The American Scholar* in the *Bulletin of the Atomic Scientists,* August, 1947, and the several comments on Ridenour's article by a number of American scientists and scholars.

21. J. B. Conant, "Science and politics."

22. J. P. Baxter, *Scientists Against Time,* pp. 28ff. See also, Irvin Stewart, *Organizing Scientific Research for War,* Boston: Little, Brown and Co., 1948, pp. 326ff.

23. J. R. Newman and B. S. Miller, *The Control of Atomic Energy.*

24. Henry D. Smyth, "The role of the national laboratories in atomic

energy development," *Bulletin of the Atomic Scientists,* VI(1950), p. 6.

25. See L. Kowarski, "Psychology and structure of large-scale physical research," for a discussion of the structure and what he calls the "pathology" of large-scale Government atomic research groups. Kowarski, now Technical Director of the French A.E.C., worked during the war in both the American and the Canadian atomic research laboratories.

26. H. D. Smyth, "The role of the national laboratories."

27. P.S.R.B., *A Program for the Nation,* p. 1ff.

28. *Fortune Magazine,* "The scientists," p. 166.

29. P.S.R.B., *The Federal Research Program,* p. 1ff.

30. David Lloyd Kreeger, "The control of patent rights resulting from federal research," *Law and Contemporary Problems,* XII(1947), p. 714.

31. *Ibid.*

32. *Fortune Magazine,* "The scientists," p. 176.

33. Talcott Parsons, "The science legislation," contains an analysis of the reasons for the failure of the early N. S. F. legislation.

34. Sen. H. M. Kilgore, "Science and the Government," *Science,* 102(1945), 630-638, is an early statement of the case "for" the N. S. F.

35. See the complete text of the National Science Foundation bill in *Bulletin of the Atomic Scientists,* VI (1950), 186-190.

CHAPTER IX

1. S. C. Gilfillan, *The Sociology of Invention,* Chicago: Follett Publishing Co., 1935; S. C. Gilfillan, *Inventing the Ship,* Chicago: Follett Publishing Co., 1935; and Joseph Rossman, *The Psychology of the Inventor,* Washington, D.C.: Inventors Publishing Co., 1931.

2. R. Dubos, *Louis Pasteur,* p. 374.

3. D. H. Killeffer, *The Genius of Industrial Research,* p. 6.

4. M. Polanyi, *Science, Faith and Society,* p. 29. Further on the virtues of intimate contact among scientists, see L. Infeld, *Quest,* p. 285, where he tells how scientists laughed when Einstein said that the ideal job for a scientist would be that of lighthouse keeper.

5. H. Levy, *The Universe of Science,* p. 195. Further on the use of metaphor in scientific imagination, see Eduard Farber, "Chemical discoveries by means of analogies," *ISIS,* 41(1950), 20-26, for an extensive account of many chemical discoveries based on the use of analogy; and David Lindsay Watson, *Scientists are Human,* London: Watts & Co., 1938, Ch. VI, "On the Similarity of Forms and Ideas—As the Basis of Mental Life and of Science."

6. Ralph C. Epstein, "Industrial invention: heroic or systematic?" *Quarterly Journal of Economics,* XI(1925-26), p. 241.

7. Edwin H. Sutherland, *White Collar Crime,* New York: The Dryden Press, 1949, p. 105.

8. Norman T. Ball, "Research or available knowledge: a matter of classification," *Science,* 105(1947), p. 34.

9. T. A. Boyd, *Research,* p. 78.

10. John C. Stedman, "Invention and public policy," *Law and Contemporary Problems,* XII(1947), p. 664.

11. *Ibid.* Further on the problem of novelty in patentable inventions, see J. Harold Byers, "Criteria of patentability," *The Scientific Monthly,* 62(1946), 435-439.

12. George Sarton, *The History of Science,* p. 16.

13. Lewis Mumford, *Technics and Civilization,* New York: Harcourt, Brace and Co., 1934, p. 142.

14. J. A. Hobson, *The Evolution of Modern Capitalism,* rev. ed., p. 79, cited in R. C. Epstein, "Industrial invention," p. 243.

15. S. C. Gilfillan, *Inventing the Ship.*

16. On Newton, see H. Butterfield, *The Origins,* p. 136; on Einstein, see Louis deBroglie, "A General Survey of the Scientific Work of Albert Einstein," in P. A. Schilpp, ed., *Albert Einstein, Philosopher-Scientist,* Evanston, Ill.: Library of Living Philosophers, 1949, p. 114.

17. For the "heroic" view in its full nineteenth century form, see Samuel Smiles, *Lives of the Inventors.*

18. Eric Ashby, *Scientist in Russia,* pp. 196ff.

19. William Fielding Ogburn, *Social Change,* New York: B. W. Huebsch, 1922, pp. 90-122.

20. Bernhard J. Stern, *Social Factors in Medical Progress,* New York: Columbia University Press, 1927.

21. J. P. Baxter, *Scientists Against Time,* p. 139.

22. J. Rossman, *The Psychology of the Inventor,* Ch. VIII.

23. *Ibid.,* p. 132.

24. Lancelot Law Whyte, "Simultaneous discovery," *Harper's Magazine,* February, 1950, p. 25.

25. For some other studies of social influences on invention, see W. R. Maclaurin, *Invention and Innovation in the Radio Industry;* Sir Josiah Stamp, *Some Economic Factors in Modern Life,* London: P. S. King and Son, 1929; and A. L. Kroeber, *Anthropology,* new ed., rev., New York: Harcourt, Brace and Co., 1948, Ch. 11, where one may find some histories of inventions in which the interplay of many different social factors is suggested. Among the inventions Kroeber treats are mills, mechanical clocks, photography, the telephone, the zero concept, and the fire piston.

26. For a classic statement, see Claude Bernard, *An Introduction to the Study of Experimental Medicine,* and especially the introduction by L. J. Henderson. See also, René Dubos, *Louis Pasteur,* Ch. XIII.

27. Claude Bernard, *An Introduction to the Study of Experimental Medicine,* p. 155.

28. For many examples, see W. Platt and R. A. Baker, "The relation of the scientific 'hunch' to research," *Journal of Chemical Education,* VIII(1931), 1969-2002; and Jacques Hadamard, *The Psychology of Invention in the Mathematical Field,* Princeton: Princeton University Press, 1945.

29. W. B. Cannon, *The Way of an Investigator,* p. 68.

30. *Ibid.,* p. 74.

31. Among others, see Claude Bernard, *An Introduction to the Study of Experimental Medicine,* Part Three, Ch. I; W. B. Cannon, *The Way of An Investigator,* Ch. VI, "Gains from Serendipity;" E. Mach, "On the part played by accident in invention and discovery," *The Monist,* 6(1896), 161-175; René Dubos, *Louis Pasteur,* pp. 91-92, 106; J. B. Conant, *On Understanding Science,* pp. 65ff., "The Role of Accidental Discovery;" I. Bernard Cohen, *Science, Servant of Man,* Boston: Little, Brown and Co., 1948, Ch. 3; T. A. Boyd, *Research,* Ch. IX, "Accident;" J. Rossman, *The Psychology of the Inventor,* Ch. VII; Franklin C. McLean, "The happy accident," *The Scientific Monthly,* 53(1941), 61-70; P. Bridgman, *Reflections of a Physicist;* John R. Baker, *The Scientific Life,* London: George Allen & Unwin, 1942, Chs. 4, 5; and R. K. Merton, *Social Theory and Social Structure,* pp. 98-102, 375, 376-77, for cases in social science.

32. E. Mach, "On the part played by accident," p. 168.

33. F. R. Bichowsky, *Industrial Research,* p. 44.

34. *Ibid.;* see also pp. 45ff. for another case of how Bichowsky missed an invention. We are in Bichowsky's debt for the neon case especially because such "failures" of imagination are sel-

dom reported in print so clearly and in such detail.

35. R. Dubos, *Louis Pasteur,* pp. 377ff. See also, on what he calls the "illogical" progress of science, P. W. Bridgman, "Impertinent reflections on history of science," *Philosophy of Science,* 17(1950), pp. 69ff.

CHAPTER X

1. *Fortune Magazine,* "The scientists."

2. Sumner H. Slichter, "Some economic consequences of science," *The Annals,* 249(1947), p. 105.

3. Talcott Parsons, *Essays in Sociological Theory, Pure and Applied,* Glencoe, Ill.: The Free Press, 1949, pp. 266-68.

4. Thorstein Veblen, *The Instinct of Workmanship,* New York: The Macmillian Co., 1914, p. 314. See also, for the most developed statement of this theory and its corollary, the theory of "cultural lag," W. F. Ogburn, *Social Change.*

5. Bernhard J. Stern, "Resistances to the Adoption of Technological Invention," in *Technological Trends and National Policy,* Washington, D.C.: U. S. Government Printing Office, 1937; and Vincent Heath Whitney, "Resistance to innovation: the case of atomic power," *American Journal of Sociology,* LVI (1950), 247-254.

6. Bernhard J. Stern, "Restraints upon the utilization of inventions," *The Annals,* 200(1938), November, p. 14.

7. R. S. Flanders, *The Atlantic Monthly,* January, 1951.

8. Richard H. Shryock, *The Development of Modern Medicine,* Philadelphia: The University of Penna. Press, 1936, Ch. III. See also, R. H. Shryock, *American Medical Research,* New York: The Commonwealth Fund, 1947.

9. A. C. Ivy, "The history and ethics of the use of human subjects in medical experiments," *Science,* 108(1948), 1-5.

10. B. J. Stern, "Restraints," pp. 16-22. For a discussion of the opposition to invention by early capitalists who were granted royal patents of monopoly, see G. N. Clark, "Early capitalism and invention," *Economic History Review,* VI(1936), pp. 149-153.

11. Frank Joseph Kottke, *Electrical Technology and the Public Interest,* Washington, D.C.: American Council on Public Affairs, 1944, esp. pp. 58, 63. See also, W. R. Maclaurin, *Invention and Innovation in the Radio Industry;* and W. R. Maclaurin, "Patents and technical progress; a study of television," *The Journal of Political Economy,* LVIII(1950), 142-157.

12. William B. Bennett, *The American Patent System, An Economic Interpretation,* Baton Rouge: Louisiana State University Press, 1943, p. 188.

13. On the economic problems of capital cost and obsolescence with regard to inventions, see Sir Josiah Stamp, *The Science of Social Adjustment,* London: Macmillian and Co., 1937, pp. 34ff., and Yale Brozen, *Social Implications of Technological Change,* mimeographed, available from the Social Science Research Council, New York, 1950, Chs. 5, 6.

14. Alexander Morrow, "The suppression of patents," *The American Scholar,* 14(1945), p. 219.

15. Oswald Spengler, *Man and Technics,* New York: A. A. Knopf, 1932, p. 198.

16. For instances in the early seventeenth and eighteenth centuries, see G. N. Clark, "Early capitalism and invention," and Willard L. Thorp, "Evolution of Industry and Organization of Labor," in Jess E. Thornton, ed., *Science and Social Change,* Washington, D.C.: The Brookings Institution, 1939. For a recent case, see Elizabeth Baker, *Displacement of Men by Machines: Effects of Technological Change in Commercial Printing,* New York, 1933.

17. J. R. McCulloch, ed., *Works of David Ricardo,* London, 1846, p. 239. For further evidence in more recent studies, see Ewan Clague and W. J. Couper, *The Readjustment of Industrial Workers Displaced by Two Plant Shutdowns,* New Haven: Yale University Press, 1934; David Weintraub, "Unemployment and Increasing Productivity," in U. S. National Resources Committee, *Technological Trends and National Policy,* Washington, D.C.: Government Printing Office, 1937; and Nathan Belfer, "Implications of capital-saving inventions," *Social Research,* 16 (1949), 353-65.

18. On the "automatic factory," see *Fortune Magazine,* "The automatic factory," 34(1946), Nov., 160ff. and E. W. Leaver and J. J. Brown, "Machines without men," *Fortune Magazine,* 34(1946), Nov., 165ff.

19. See E. Wight Bakke, *Citizens Without Work,* New Haven: Yale University Press, 1940, and E. Wight Bakke, *The Unemployed Worker,* New Haven: Yale University Press, 1940.

20. See Elliott Dunlap Smith, *Technology and Labor,* for a study of the effects of labor-saving inventions and labor-saving management technology upon workers in the cotton textile industry. See also Robert K. Merton, "The machine, the worker, and the engineer," *Science,* 105(1947), 79-84, for the effects of the machine upon what Merton calls "the social anatomy of the job."

21. In this work William F. Ogburn has been the leader and the inspiration for a group of other men. The chief statement of this group may be found in U. S. National Resources Committee, *Technological Trends and National Policy.* For a critique of the predictions made in this book, see S. Lilley, "Can prediction become a science?", *DISCOVERY,* Nov., 1946, 336-40, where he says, "the proportion of definite positive successes is not very high—certainly not high enough to be very useful in making social and political decisions." See also W. F. Ogburn, "National Policy and Technology," in U. S. National Resources Committee, *Technological Trends and National Policy,* and W. F. Ogburn, *The Social Effects of Aviation,* Boston: Houghton Mifflin Co., 1946. See S. C. Gilfillan, "The Prediction of Inventions," in U. S. National Resources Committee, *Technological Trends and National Policy,* for an account of the long pre-history of the attempts at prediction of invention.

22. S. Lilley, "Can prediction become a science?".

23. Frederick Soddy, *Science and Life,* New York: E. P. Dutton and Co., 1920, p. 35.

24. Farrington Daniels, "Science as a Social Influence," in R. C. Stauffer, ed., *Science and Civilization,* Madison: University of Wisconsin Press, 1949, p. 166.

25. S. Lilley, "Can prediction become a science?".

26. U. S. National Resources Committee, *Technological Trends and National Policy,* p. 61.

27. W. F. Ogburn, *The Social Effects of Aviation.*

28. In this connection, note Lilley's remark: "Of course the main reason why *Technological Trends* failed in its prediction was that it failed to foresee that the depression of the 'thirties' would be transformed into the war of the early 'forties' and that the war would greatly decrease the influence of the retarding forces," in S. Lilley, "Can prediction become a science?".

29. J. R. Newman and B. S. Miller, *The Control of Atomic Energy,* p. 5.

30. R. K. Merton, *Social Theory and Social Structure,* Ch. VII.

31. Sir Frederick Soddy, et al., *The Frustration of Science,* New York: W. W. Norton & Co., 1935. "The frustration of science," says J. D. Bernal, one of the leaders of the group, "is a bitter thing," in his *The Social*

Functions of Science, p. xv. This book, especially in Ch. VI, is a typical statement of the views of the group.

32. See, for example, T. Swann Harding, *The Degradation of Science,* New York: Farrar & Rinehart, 1931.

33. On the general tendency of experts to absolutize their technical ideals, see F. A. von Hayek, *The Road to Serfdom,* Chicago: The University of Chicago Press, 1944, p. 53. Von Hayek says, "Almost every one of the technical ideals of our experts could be realized within a comparatively short time if to achieve them were made the sole aim of humanity." Also, on this point, see Sir Josiah Stamp, *The Science of Social Adjustment,* p. 50.

34. K. Davis, *Human Society,* p. 438.

35. Norbert Wiener, "A scientist rebels," *The Atlantic Monthly,* 179 (1947), p. 46.

36. Louis N. Ridenour, "The scientist fights for peace," *The Atlantic Monthly,* 179(1947), 80-83.

37. P. W. Bridgman, "How far can scientific method determine the ends for which scientific discoveries are used?".

38. See J. B. S. Haldane, *Callinicus: A Defence of Chemical Warfare,* London: Kegan Paul, 1924, and J. B. S. Haldane, *Daedalus, or Science and the Future,* New York: E. P. Dutton & Co., 1924.

39. L. N. Ridenour, "The scientist fights for peace."

40. P. W. Bridgman, "How far can scientific method determine the ends for which scientific discoveries are used?".

41. Edward A. Shils, "A critique of planning—The Society for Freedom in Science," *Bulletin of the Atomic Scientists,* 3(1947), no. 3. The organizational locus of the British "planners" remains in the Association of Scientific Workers, which in 1949 had some 20,000 members, including engineers and technicians as well as professional scientists. See J. D. Bernal, *The Freedom of Necessity,* p. 18.

42. J. D. Bernal, *The Freedom of Necessity.*

43. Pamphlet issued by the Society, May, 1944.

44. J. D. Bernal, *The Social Functions of Science,* p. 329.

45. M. Polanyi, "The case for individualism," *Bulletin of the Atomic Scientists,* 5(1949), no. 1.

46. J. D. Bernal, *The Social Functions of Science,* p. 325.

47. S. I. Vavilov, *Soviet Science: Thirty Years,* Moscow: Foreign Languages Publishing House, 1948, p. 33. In general, this whole article by Vavilov is far from doctrinaire on the nature of science and on the possibilities of "planning." On the concrete details of "planning" in Russian science, see Eric Ashby, *Scientist in Russia, passim.*

48. See M. Polanyi in E. P. Wigner, ed., *Physical Science and Human Values,* p. 126, 129. See also, Frank H. Knight, "Virtue and knowledge: the view of Professor Polanyi," *Ethics,* LIX(1949), 271-284, for a critique of Polanyi's more extreme statements.

CHAPTER XI

1. Elton Mayo, *The Human Problems of an Industrial Civilization,* New York: The Macmillan Co., 1933, and F. J. Roethlisberger and W. J. Dickson, *Management and the Worker,* Cambridge: Harvard University Press, 1939.

2. Malcolm G. Preston, "Concerning an essential condition of cooperative work," *Philosophy of Science,* 15 (1948), p. 97.

3. See K. Davis, *Human Society,* Part V, for some good examples of this kind of work.

4. For a brief attempt to account for the sources of social science in the social thought of the eighteenth century, see Bert F. Hoselitz, "The social

sciences in the last two hundred years," *The Journal of General Education,* IV (1950), 85-103.

5. See, e.g., Daniel Lerner and Harold D. Lasswell, eds., *The Policy Sciences,* Palo Alto: Stanford University Press, 1951, Part II.

6. Talcott Parsons, *The Social System;* R. K. Merton, *Social Theory and Social Structure.*

7. *Scientific American,* July, 1950, p. 13. See also, Elbridge Sibley, *The Recruitment, Selection, and Training of Social Scientists,* New York: The Social Science Research Council, 1948.

8. George A. Lundberg, "The Senate ponders social science," *The Scientific Monthly,* 64(1947), 397-411.

9. Testimony before a subcommittee of the Committee on Interstate and Foreign Commerce, House of Representatives, 79th Congress, second session, May 28, 29, 1946. Washington, D.C.: Government Printing Office, 1946, pp. 11, 13.

10. *Fortune Magazine,* "The scientists," p. 176. For a similar favorable expression by natural scientists, see the results of the survey conducted by the Inter-Society Committee for a National Science Foundation among 70-odd scientific and educational organizations, in *Science,* 105(1947) p. 329.

11. J. B. Conant, "The scientist in our unique society," p. 49.

12. Julian Huxley, "Science, Natural and Social," in R. N. Anshen, ed., *Science and Man,* New York: Harcourt, Brace and Co., 1942, p. 286.

13. See A. M. Schlesinger, Jr., "The statistical soldier," *Partisan Review,* Summer, 1949; and Daniel Lerner, " 'The American Soldier' and the Public," in Robert K. Merton and Paul F. Lazarsfeld, eds., *Continuities in Social Research,* Glencoe, Ill.: The Free Press, 1950.

14. See accounts of these in U. S. National Resources Committee, I. *Relation of the Federal Government to Research,* pp. 133ff.

15. John McDiarmid, "The Mobilization of Social Scientists," in Leonard D. White, ed., *Civil Service in Wartime,* Chicago: University of Chicago Press, 1945.

16. Alexander H. Leighton, *Human Relations in a Changing World,* New York: E. P. Dutton & Co., 1949.

17. Roy R. Grinker and John P. Spiegel, *Men Under Stress,* Philadelphia: Blakiston, 1945.

18. S. A. Stouffer, *et al., The American Soldier.* I. *Adjustment During Army Life;* II. *Combat and Its Aftermath,* Princeton: Princeton University Press, 1949.

19. The Russell Sage Foundation, *Effective Use of Social Science Research in the Federal Services,* New York: The Russell Sage Foundation, 1950.

20. U. S. National Resources Planning Board, III. *Business Research.*

21. See James C. Worthy, "Organizational structure and employee morale," *American Sociological Review,* 15 (1950), 169-179, for a study of social research in Sears, Roebuck, and Co.

22. U. S. National Resources Planning Board, III. *Business Research,* p. 4.

23. Solomon Barkin, "Applied social science in the American trade-union movement," *Philosophy of Science,* 16(1949), 193-197.

24. Robert K. Merton, "Role of the intellectual in public bureaucracy," *Social Forces,* 23(1945), 405-415; Robert K. Merton, "The role of applied social science in the formation of policy: a research memorandum," *Philosophy of Science,* 16(1949), 161-181; E. A. Shils, "Social science and social policy," *Philosophy of Science,* 16(1949), 219-242; and A. H. Leighton, *Human Relations in a Changing World,* pp. 129-146.

25. R. K. Merton, "Role of the intellectual," p. 407.

26. Alfred McClung Lee, "Implementation of opinion survey standards," *Public Opinion Quarterly,* 13(1950), 645-52.

27. Lee R. Steiner, *Where Do People Take Their Troubles?,* Boston: Houghton, Mifflin Co., 1945.

28. A. H. Leighton, *Human Relations in a Changing World,* p. 206.

29. For a critique of the positivist social science view, see Hans J. Morgenthau, *Scientific Man and Power Politics,* Chicago: University of Chicago Press, 1946.

Index

INDEX